INTERPERSONAL LIVING

A Skills/Contract Approach to Human-Relations Training in Groups

OTHER BOOKS BY GERARD EGAN

ENCOUNTER
Group Processes for Interpersonal Growth

FACE TO FACE
The Small-Group Experience and Interpersonal Growth

THE SKILLED HELPER
A Model for Systematic Helping and Interpersonal Relating

EXERCISES IN HELPING SKILLS
A Training Manual to Accompany The Skilled Helper

INTERPERSONAL LIVING

A Skills/Contract Approach to Human-Relations Training in Groups

GERARD EGAN
Loyola University of Chicago

Brooks/Cole Publishing Company
Monterey, California

A Division of Wadsworth Publishing Company, Inc.

ISBN: 0-8185-0189-8
L.C. Catalog Card No.: 76-6651
Printed in the United States of America

10 9 8 7 6 5 4

Production Editor: *Konrad Kerst*
Interior & Cover Design: *John Edeen*
Typesetting: *Dharma Press, Emeryville, California*
Printing & Binding: *R. R. Donnelley & Sons Company, Crawfordsville, Indiana*

Preface

This book is meant for anyone who wishes to improve his or her interpersonal skills. It should be of use to anyone in a people-oriented occupation. It is especially relevant for those who want to improve their interpersonal skills or interactional style through some kind of group process. And, since the skills discussed in this book are basic interpersonal skills, this book may be used in courses designed to improve interpersonal communication, whether in one-to-one or group situations.

In a book I published with Brooks/Cole in 1970 (*Encounter: Group Processes for Interpersonal Growth*) I noted that, despite the fact that all sorts of group experiences designed to promote personal and interpersonal growth were being conducted, and that thousands of people were flocking to such groups, comparatively little had yet been written on these groups in terms of either theory or research. Since then, however, there has been an explosion of publications dealing with such groups. Today, therefore, not only is it bewildering to behold the different kinds of group experiences being offered, but it is an awesome task to stay abreast of the literature on these groups. You may well ask "Why, then, another book dealing with group processes?" The answer lies in two words—*skills* and *systematic*.

First, this book deals with *skills.* Much of the current literature on groups overlooks the fact that interacting in a group is a high-level communication experience requiring relatively sophisticated interpersonal-communication skills. Participants in various group experiences are often asked to engage in interactions requiring communication skills they simply don't possess. Therefore, this book takes a step backward, as it were, and deals with basics—that is, the kinds of basic interpersonal skills one needs to perform effectively in *all* interpersonal situations (and especially in groups). This book also deals with a group of basic skills that are *specific to groups*. These group-specific communication skills are extremely important, but to the best of my knowledge they are not clearly identified and illustrated in the current group literature. It is my conviction that this ignoring of basic skills—whether interpersonal-communication skills or group-specific skills—has been detrimental to the use of groups to promote personal and interpersonal growth.

The second word is *systematic*. Many group processes expose the trainee to a variety of experiences, assuming that mere exposure will enable him or her to acquire an adequate level of interpersonal skills. For instance, it seems to be assumed that a participant in a group in which a great deal of confrontation goes on will pick up the skill of responsible confrontation. Research does not support this assumption. Systematic training in specific communication skills—training that includes cognitive input and practice—offers a more certain approach to acquiring these skills. This book emphasizes systematic training in both individual and group-specific communication skills. I don't suggest that systematic training in skills provides the complete answer to the problems of interpersonal living or of interpersonal growth, but it does provide the tools that make interpersonal growth possible and interpersonal living effective.

It is difficult to thank everyone specifically who has contributed to the fashioning of this book. I do want to thank my colleague Mike O'Brien for his constant support and express special gratitude to Maureen Bacchi, Kevin O'Keefe, and Jim Armstrong for their special assistance.

I would also like to thank the reviewers of the manuscript for their helpful suggestions: John C. Clem of New River Community College, Gloriann B. Koenig of DeKalb Community College, Jim Ludlow of Modesto Junior College, C. K. Simpson of The Cleveland State University, and Jerry Wesson of El Centro College.

Gerard Egan

Contents

EXERCISES

On Interpersonal Style

The Log

On Self-Disclosure

On Concreteness

On Expressing Feelings and Emotions

On Attending

On the Communication of Accurate Empathy

Exercises Encouraging Group-Specific Skills

On Flight

Open-Group Exercises

PART 1

Introduction and Overview

Chapter 1

Introduction: The Basic Concepts of the Human-Relations-Training Laboratory

DEFINING THE HUMAN-RELATIONS-TRAINING EXPERIENCE

Since there are at least dozens of different kinds of laboratory experiences that deal with human-relations training, the development of human potential, self-actualization, group dynamics, and the like, it is essential to give some kind of definition to the experience you are about to enter. Perhaps you have heard about "sensitivity training" or "encounter groups," and perhaps you have even participated in one or two group experiences. "Sensitivity training" has come to mean so many different things that it means practically nothing today. And, although I have used the term "encounter group" in earlier writings (Egan, 1970, 1971b, 1973a), I have abandoned it because it, too, now has too many different connotations. Therefore, it seems more reasonable to try to define the human-relations-training experience simply as it is developed in this book, without trying to connect it historically to all of the group experiences and laboratories that have proliferated over the past ten years.

You may well be fearful of the prospect of entering a "human-relations-training" experience, but learning something about what awaits you may help to dispel some of this fear. Actually, "awaits you" is a poor phrase, for human-relations training is something you *do*, not something done to you. This chapter later offers an overview of precisely what you are to do during each of the three phases of this training program. The introductory sections leading up to this overview will define such terms as "human-relations training," "laboratory experience," and "interpersonal style" in order to make the overview as clear as possible. The more clearly you understand the experience before you embark upon it, the more intelligently you will be able to give yourself to it and the more valuable it will be to you.

HUMAN-RELATIONS TRAINING

In order to act effectively in interpersonal situations, ideally you should

- know what you are like interpersonally—how you think, feel, and act in characteristic interpersonal situations and how others characteristically think, feel, and act toward you—that is, have some understanding of your "interpersonal style" (this term is explained more carefully later in this chapter);
- have whatever interpersonal skills you need to relate to others effectively, both individually and in groups;
- be willing to change those aspects of your interpersonal style that are unproductive and that cause you and others trouble;
- know in what ways you would like to grow interpersonally; and
- know and practice the values you would like to let govern your interpersonal life.

The human-relations-training experience outlined in this book is designed to help you achieve the goals listed above. It provides

- an opportunity for you to come to a deeper understanding of your own interpersonal style;
- a chance for you to review your level of competence in interpersonal skills;
- the training you need to acquire and/or develop sets of core interpersonal skills;
- an opportunity to change the way you act interpersonally in ways you consider growthful for yourself (including getting rid of problematic behaviors and developing more growthful behaviors);
- feedback from others as you go through this process of change; and
- an opportunity to examine the values that govern your interpersonal life in the community of your small group.

SOME ELEMENTS OF LABORATORY LEARNING

Human-relations-training experiences and other related small-group experiences that focus on personal and interpersonal growth are often referred to as "laboratories." Courses in colleges and universities often use the word "laboratory" in describing human-relations training or training in communication skills—for instance, "Interpersonal Relations: A Laboratory Approach" or "A Laboratory in Communication Skills." Since *Interpersonal Living* is a laboratory approach to learning, I will first describe some of the major features of this kind of learning.

Experiential learning. Although a certain amount of your learning will take place through reading, lectures, and written exercises, your most important learning will come from your actual training in and use of interpersonal skills in face-to-face conversation with your fellow group members. Theory is subordinated to involvement; indeed, the whole purpose of the laboratory is to translate theory into action. For instance, after you have read a few things about self-disclosure, you will engage in self-disclosure. In general you will learn by doing.

The small group. You will be a member of a relatively small group. At times, this group will be divided into even smaller units (into threes, for instance, in order to practice skills). You will remain in this group throughout the laboratory experience. It will become your community, the center of your learning.

Feedback. As you learn skills and interact with your fellow group members, you will both observe the behavior of others and be observed by them. Thus you will all have a rich opportunity to express how you are affected by one another's behavior. Constructive feedback from group members who are committed to one another's growth is one of the most exciting possibilities in laboratory learning. This resource is lacking in learning situations that are highly (or overly) cognitive.

A climate of experimentation. If the group experience is to have impact on your behavior outside the laboratory, if it is to make a difference in day-to-day interpersonal living, then the laboratory experience must be *different* from day-to-day experience. For this reason, you are encouraged to experiment with your behavior during the laboratory—that is, to attempt constructive behaviors that up to this point have not been part of your interpersonal patterns. For instance, if you never confront anyone in your day-to-day interpersonal life, you might benefit from exercising the skill of confrontation in the lab. Laboratories generally provide participants with a kind of "cultural permission" to engage in forms of responsible behavior that the constrictions of the back-home environment do not allow. This does not mean that the laboratory is a setting for "wild" or irresponsible behavior, but freeing yourself of unproductive restrictions in your interpersonal transactions is hardly irresponsible.

Structure. Laboratories often provide participants with some kind of structure in order to help them channel their energies toward achieving specific goals. In a study dealing with structured and unstructured human-relations training, Levin and Kurtz (1974) found that participants in structured groups reported greater ego involvement in their group, more self-perceived personality change since joining the group, and greater group unity than did participants in unstructured groups.

The laboratory experience described in these pages has, at least relatively speaking, a high degree of structure. I have dealt with the questions of structure and of high visibility and goal clarity versus low visibility and goallessness elsewhere (Egan, 1970, 1971a, 1971b, 1973a, 1973b, 1975a, 1975b; also see Carkhuff, 1971; Ivey, 1971; and Kagan, 1971, for related discussions), and this discussion need not be repeated here. It is my belief that systematic approaches to human-relations training, including interpersonal-skills training, are more effective than goalless, scattershot approaches. When the structure is highly visible, nonmanipulative, and facilitating, the fears held by some humanistic psychologists and educators that systematic approaches are controlling and dehumanizing are unfounded.

Bednar, Melnick, and Kaul (1974) suggest that lack of structure in early group sessions tends to accentuate participant fear and leads to unrealistic expectations. According to these researchers, a group will be more effective if the participants have a clear understanding of such factors as group goals, the process of group development, what the members expect of the group leader or leaders, and what the leader or leaders expect of the participants. These authors have found that relatively firm structure in the beginning of a group experience enables the participants to begin vigorously—that is, to engage in the kinds of risk-taking that are both reasonable and appropriate to group goals. The problem with structure, they say, is that too much or too prolonged structure keeps the participants from exercising their own initiative. The structure presented in this book is meant to help you, as the Roman poet Horace says, get "into the middle of things" as quickly as possible. As you acquire skills and become more experientially familiar with the core contract of the laboratory, structure is reduced to enable you to exercise more and more initiative. (For further discussions on structure, especially initial structure, see Crews, 1974; Lee, 1974; Melnick, 1974; and Roach, 1974.) Chapter 2 outlines the specific structure underlying the *Interpersonal Living* laboratory and the "contract" at the heart of that structure.

The here and now. Laboratories generally deal with what is happening here and now in the group. Little energy is invested in recalling data from any member's interpersonal past, and little time is spent on what takes place outside the group. Perhaps a better way of putting it is that you can certainly deal with your past or with what is going on outside the group *provided* that you relate there-and-then material to what is happening here and now in the group. For instance, a person might say "Whenever my wife or children make demands on me, I grow silent and tend to spend less time with them. Now that you people are beginning to place legitimate demands on me, I notice the same thing happening here; I can't withdraw physically, but I notice that I can withdraw psychologically." In this instance, a there-and-then concern is related immediately to what is happening in the group. Since a major part of your work in the group will be concentrated on establishing and developing relationships with your fellow group members, it is only natural that the conversation revolve around the here and now.

Leadership. In laboratory experiences a leader is usually designated for each small group. This person is called the "facilitator" or "trainer." In the

Interpersonal Living lab, the trainer is actually both "leader" and "member." He or she will provide the structure and the direction in the skills-training sessions that constitute part of the lab. Later on in the group experience, he will model the kinds of behavior that make groups high-level experiences, and he will encourage you and the other members to implement the overriding "contract" of the group (a concept that will be explained in Chapter 2). Ideally, leadership will become diffused within the group; that is, the members, once they acquire basic interpersonal skills and become familiar with the contract, will take the initiative to involve themselves with their fellow members without having to be prompted to do so by the leader. As the group moves on, the leader should become more and more a good group member and less and less a leader. Overdependency on the leader usually suffocates the life of the group.

Anxiety. Do you feel somewhat anxious about entering into this experience? If you do, welcome to the club—because most people do. Participation in the group is seductive, for it is an approach to the unknown and usually offers some hope for self-knowledge and self-improvement. But it is also anxiety arousing, for even though participation in the group offers a fresh source of relatedness it also demands self-investment, and one's self-image is at stake. "What if I flop?" is on the minds of many prospective participants. The Yerkes-Dodson Law (Yerkes & Dodson, 1908) suggests that an anxiety level that is either too high or too low is not good for learning. A reasonable amount of anxiety acts as an energizer. The contract that governs the laboratory you are about to enter will give you some idea of what is expected of you and how you are to invest yourself. If this contract arouses some anxiety in you, you will not be alone, and you will have an opportunity to express and deal with your anxiety at the very beginning of the group experience.

People differ greatly in both "state" anxiety (how much anxiety they experience in a given anxiety-arousing situation) and in "trait" anxiety (anxiety that is more or less constant, anxiety that forms part of a person's personality makeup; see Levitt, 1967). Therefore, you may be more anxious or less anxious at any given moment than your fellow participants. The laboratory, however, will give you an opportunity to deal with your anxiety, especially the anxiety that comes from your need to achieve during the lab (see Lawrence & Lorsch, 1969; McClelland, 1961) and the anxiety that comes from the need for affiliation—that is, more strictly interpersonal anxiety (see Atkinson, 1958).

Psychological safety. Whereas laboratories offer possibilities for risk-taking, they should also provide a climate of psychological safety that makes such risk-taking possible. The structure of the *Interpersonal Living* laboratory attempts to induce this climate in various ways. It adds high visibility to the training experience, and it is antimanipulative. This structure helps create a common group culture, so that the risks you take are taken in the context of this shared culture. It stresses the constructive dimensions of all skills (for instance, confrontation that is caring rather than punitive). Finally, group leaders both model constructive interpersonal behavior and encourage such behavior in others.

Cultural permission. "Cultural permission" means that participants in the laboratory experience certain kinds of freedom not available in the day-to-day culture in which they live. For instance, one such "permission" is that relative strangers (the lab participants) feel freer to talk about themselves at comparatively deep levels than they ordinarily feel in their everyday lives. In a sense, the cultural prerequisites for friendship and intimacy are laid aside, and the participants deal with one another at some depth, not because they are longtime acquaintances but merely because they are fellow human beings. Feedback (as described above) is another one of the cultural permissions. We seldom discuss with others the impact we have on one another. In fact, in our culture it seems more permissible to tell a third party our impressions of another that we would not dare tell the other. And, although such conversations may satisfy a need to vent one's frustrations in interpersonal living, there is little satisfaction in them that could be called growthful. As you read the overview section in Chapter 2 and mull over what you are contracting for in this group, perhaps you can identify further "cultural permissions" that seem to be afforded by this lab. There has been some complaint that certain laboratories go too far, that they permit or encourage too much, so that the cultural and even the ethical sensibilities of some of the participants are offended. However, this laboratory experience encourages responsible cultural permissiveness without subscribing to cultural or ethical license. It is hoped that you will feel freer, as you move through the laboratory experience, to experiment responsibly with dimensions of your interpersonal style that you do not feel so free to experiment with in everyday life.

Artificiality and reality in laboratory life. At first glance, a laboratory in interpersonal relations appears to suffer from a relatively high degree of artificiality. In a sense this is true. Laboratories are by definition contrived, and their artificiality would make little sense unless it had an impact on day-to-day living. For instance, the people who constitute the various groups in the laboratory situation do not come together naturally; you do not get to be with those to whom you are naturally attracted. You are assigned to a group, and the people you find yourself with are, for the most part, strangers. And yet the culture of the laboratory demands that you achieve a certain kind of intimacy with one another in the give and take of group interaction. The structural elements of the laboratory—for instance, skills training and other kinds of exercises—also contribute to the artificiality of the laboratory. Not only are you with strangers, but the ways in which you are to relate to them are defined, to a degree, by these structural elements and by the overriding contract that gives definition to the laboratory.

On the other hand, the artificiality of the laboratory can help you face many of the realities, perhaps especially the overlooked realities, of interpersonal living. For instance, we live in a society of high and increasing mobility. Although Aristotle suggested that people cannot become friends in the truest sense of the term until they have eaten a "peck of salt" together, in modern society the ability to establish relationships of some meaning and closeness relatively quickly is essential if one is to avoid isolation. And although the laboratory does not espouse "instant intimacy" as a cultural value, it does allow you the opportunity to discover that it is possible to make meaningful human contact with strangers in a relatively short time.

You involve yourself with the members of your group not because they are strangers but because they are human beings. By joining the laboratory experience freely, you commit yourself to a particular group of people and to facing the interpersonal realities that develop in this group. Even though you live in a highly mobile society, it is also one of the realities of human living that you are locked to some degree into relationships with particular individuals and groups. In this laboratory, interpersonal problems cannot be solved by ignoring them, by moving to a different group, or by using other modes of interpersonal flight. The pressure for involvement with this particular set of people, although artificial in one sense, highlights the unproductive modes of involvement you use with the "real" people in your normal life situation. Moreover, dealing with the stranger in the laboratory can bring home to you, in a dramatic way, your failure to deal with the "stranger" element in those with whom you are intimate in real life. The laboratory, then, does not allow the opportunities for flight from intimacy that day-to-day living often does; at least, tendencies to flight are openly challenged. Therefore, part of the impact of the laboratory arises from its being more real, rather than less real, than ordinary interpersonal living.

The lab you are entering demands not only that you deal with a certain set of people but also that you do so in specified ways. In a successful laboratory the participant learns how to talk about himself, how to engage in self-exploration that is neither exhibitionistic nor masochistic, how to show care and concern for others, how to see the world through the eyes of others, how to foster constructive emotions and avoid or handle destructive ones—in short, how to function more fully in interpersonal living. In his "real" life, a participant may manage to avoid letting others, even friends, know about the "person inside"; he may court peace at any price in his contacts with others and thus avoid both confrontation and the self-examination that the confrontational process involves. But the pressures of the laboratory force him to face, at least to some degree, these realities of interpersonal living.

Two more sources of artificiality are found in this book—the examples and the exercises. A large number of examples are used to clarify a theory or to illustrate a skill being taught. Sometimes these examples are too pat; that is, they indicate statements or responses that might be made ideally, but often simply are not made, in the give and take of interpersonal communication. If some of the examples represent an ideal, then let them be taken as such. They are in the text as teaching aids, not as phrases that you should memorize and repeat. And, since all examples are to some degree taken out of context, they cannot escape sounding artificial at times. But examples also provide a reality missing in texts that expound pure theory, and they do put flesh on the (sometimes quite dry) bones of theory.

There are many exercises in this book. They invite you to practice in a variety of ways the skills you will be learning. Obviously, doing exercises is not the same as actually relating interpersonally, and there is always something "plastic" about writing and practicing responses. But I have found that exercises, for all their "plastic" character, do help lab participants to master and concretize theory. Exercises also constitute a kind of "rehearsal" that facilitates the use of skills in real interpersonal settings.

There is nothing mystical about examples or exercises. Use them to the degree that they help you to improve your skills.

Finally, any lab or workshop remains ultimately artificial if you cannot transfer what you learn there to your everyday life; but the transfer value of the laboratory experience outlined in this book is estimated to be high. The question of transfer value will be treated more directly in the last chapter of the book.

These, then, are some of the elements common to a wide variety of laboratory-learning experiences. It is impossible for you to sort out and understand all of these elements immediately. You will be able to do this only with some experience in the group itself. However, the brief discussion of them in this chapter will prevent your going into the group experience "cold."

D-NEEDS, B-NEEDS, AND M-NEEDS

One of the steps necessary in creating a climate of psychological safety is to be aware of the dynamics involved in an individual's needs. Understanding these growth needs will help you to contribute more to your group experience.

Maslow (1968) sees the origin of neurosis in a person's being deprived of certain satisfactions that he calls needs, in the same sense that water and amino acids are needs: their absence produces illness or maladjustment. Some of these basic needs are for belongingness and identification, for close love relationships, and for respect and prestige. These are called D-needs (D for deficiency), for such needs, if unfulfilled, stand in the way of further human growth. Another set of needs are called B-needs (B for being). These are the needs of a person whose D-needs have been adequately satisfied but who still feels a drive for further self-actualization. For instance, a person whose basic needs have been met will probably have a need for B-love rather than for D-love. D-love is selfish love; it is characterized by possessiveness, anxiety, and hostility. B-love, on the other hand, is love for the being of the other person. It is unselfish, nonpossessive, capable of being constantly deepened, and characterized by minimum degrees of anxiety and hostility.

I personally believe, however, that another category—M-needs (M for maintenance)—may be added to Maslow's scheme. Although many people may not be grappling with any D-problem, they still have not moved on to any significant pursuit of B-values in key areas of life, such as the area of interpersonal relationships. Their relationships with others are not noticeably destructive, but neither are they growthful and engaging. They are bland. Their home lives are neutral—neither hotbeds of neurotic interaction nor centers of interpersonal stimulation. They profess certain religious values—values that should draw them closer to their fellow human beings—but in practice these values are ritualistic and restraining, holding them back from doing wrong rather than impelling them to involve themselves as effective helpers in their communities. Too many of us, perhaps, exhaust an excessive amount of our energies in M-functions, in "keeping life going," so that we have comparatively little energy left over for B-functions. If that is the case, we are not growing; our growth is on a plateau, immobile, and we will inevitably become the victims of our own boredom.

If laboratory groups were composed of participants with severe unresolved D-needs, then they would be much more similiar to therapy groups than to groups involved in human-relations training. Laboratory groups are usually made up of participants with a mixture of D-problems, various degrees of M-function overinvolvement, and B-aspirations and skills. D-needs are not the overriding concern of the group, but nearly all participants express some D-concerns. A more important focus, perhaps, is M-overinvolvement. It is the person who is overcommitted to M-operations in his personal and interpersonal living who is the principal victim of what Maslow (1968) calls the "psychopathology of the average." One of the functions of this lab experience is to "unfreeze" or loosen up the person who is on a plateau, especially in the area of interpersonal living. An attempt will be made to apply these D-, M-, and B-need conceptualizations to concrete examples in this book.

INTERPERSONAL VALUES, MOTIVATION, AND RESPONSIBILITY

Values imply action and therefore are to be distinguished from feelings, attitudes, ideas, and ideals. If you are asked whether helping friends when they make reasonable requests for help is a value for you, you may answer "yes, certainly!" However, when you examine your *behavior*, you discover that you seldom actually do help friends, even when their requests are reasonable. You have a previous engagement, you are too busy, you have to study, you will be away that day, you are tired—there is almost always some reason why you are not available to help. Your friends finally get the message and stop asking for your help. Thus, helping friends may seem like a good idea to you, but since the idea is not translated into behavior, you cannot claim that helping others is one of your values. We find out what our real values (as opposed to ideal or notional values) are by examining our actual behaviors—the ways in which we invest our time, money, energies, and resources.

Raths and Simon (1966) discuss some of the elements in a value. A behavior is a value if

- you prize it or esteem it (for instance, listening to others without prejudging them is something "good" in your eyes);
- you choose the behavior freely from alternatives (you are not forced to listen nonjudgmentally to others);
- you declare publicly, especially by your *actions*, that you prize this form of behavior (you actually do listen to others, and others know that this action is important to you); and
- you pursue the prized behavior consistently (others find that they can count on you to be a good listener).

Your values, then, are or are seen in your preferred behaviors; your values are your behavioral priorities.

This description of values implies, of course, that values can differ from person to person. Therefore, some of your values may not be the same as mine, and you may even see some of my values as limiting or self-defeating. For instance, if I prize comfort, pleasure, and self-gratification highly, helping others may not be much of a value for me, because it interferes with my comfort. As commonly used, however, the term "value" usually has some sort of positive connotation. Thus, if I value some form of self-gratification to the degree that it interferes with your rights, my "value" will be seen by you as a "vice." In general, values that are self-defeating or that are detrimental to others or to the community at large are called vices.

This is not the place for an extended discussion of the notion of values (for further reading on the subject, see the references at the end of this chapter), but for a number of reasons it is essential to consider the question of interpersonal values early in the lab experience. Let's examine some of them.

1. *Interpersonal values are reflected in interpersonal style.* Here are a few examples of some typical interpersonal values that affect interpersonal style:

- being safe and comfortable in all interpersonal relationships and transactions;
- sharing myself deeply with trusted friends;
- having a good time with others—interpersonal fun, pleasure, and self-gratification;
- not gossiping, in order to protect the reputation of another; keeping confidential what is told to me in secret;
- desiring interpersonal "peace at any price"; being compliant and submissive with others; being a follower; ingratiating myself with others so that they will accept me;

- having many friends and acquaintances;
- manipulating and controlling others; being dominant and aggressive; getting my way when I am with others; and
- taking pains to understand others from their point of view.

Since your interpersonal style reflects your interpersonal values, this lab experience is an opportunity for you to examine your interpersonal values.

2. *Conflicts arise from differing interpersonal values.* These conflicts may be *inter*personal or *intra*personal. For instance, I may prize both personal comfort and helping others, and I find that often enough one interferes with the other. I must struggle within myself in choosing which value is to predominate. Or, in my dealings with you, I find that I prize self-sharing at the deepest levels, whereas you prize keeping your problems and deepest thoughts to yourself. This difference causes conflict in our relationship. If I do not become aware of the fact that I am pursuing two values that conflict with each other, I will remain at odds with myself. If I do not face the fact that others have interpersonal values that differ from mine, I will remain at odds with others. Research shows that conflict is usually destructive only when it is ignored or allowed to degenerate into hostility. Conflict is part of human living and is growthful when faced and worked through.

3. *The laboratory experience presented in this book is based on certain value assumptions.* For instance, it is assumed that it is "good"

- to develop a wide repertoire of interpersonal skills;
- to communicate empathic understanding to others;
- to develop behavioral ways of manifesting respect for others;
- to be open and self-disclosing in appropriate ways with friends and even associates at work;
- to develop behavioral ways of manifesting genuineness in interpersonal transactions; and
- to challenge others to do what they say they want to do, especially in terms of their interpersonal style.

These and other assumptions will become much more concrete as you move through the laboratory experience. The point I wish to emphasize here is that the value assumptions of the laboratory may not be yours, and that such a discrepancy can lead to conflict within the laboratory itself. The purpose of the laboratory is not to force you to adopt a particular set of interpersonal values but rather to have you examine and experiment with certain interpersonal skills, behaviors, and values. It is possible that the laboratory experience will constitute for you a kind of "values education"; that is, without being "brainwashed," you may come to prize certain kinds of interpersonal behavior (for instance, greater openness, working at understanding others from their frame of reference, and so on) that up to now have not been a consistent part of your interpersonal style. If conflicts arise between you and the values proposed here or between your values and those of your fellow participants, it will be growthful to face these conflicts openly. Research has shown that the lab participant who decides from the beginning, either overtly or covertly, that the lab is not going to have any behavioral impact on him involves himself in a self-fulfilling prophecy. I hope that you and your fellow participants are open to being affected by the lab in ways consonant with your own values.

Personal Choice and Responsibility

In the process of growing up, we adopt the attitudes of and imitate the behaviors of significant adults in an unreflective way. Such behavioral patterns cannot be called values in the full sense of the term, since they are not freely chosen. But gradually we begin to reflect on our behaviors, and, once this process begins, the possibility of choosing freely begins with it. Gradually our behaviors and the values they embody become our responsibility. Those of us who have had poor or even disastrous contacts with significant adults (such as parents and teachers) while growing up may well become aware that we have been saddled with self-defeating behavioral patterns. However, the time comes when each of us realizes that he or she is responsible for his or her own behavior. I can admit that I am, to a great extent, a product of past forces beyond my control, but although I cannot change the past I *can* take charge of my own behavior and change the future. Even if it means involving myself in a painful process of emotional reeducation, I can begin to choose my own set of interpersonal values and the elements of my own interpersonal style. I cannot control the behavior of others, but I can do a great deal to master my own.

Applying these principles to the laboratory experience you are about to enter, you can choose how you are going to involve yourself with your fellow participants.

Motivations

When you ask yourself why you are involving yourself in this laboratory, what answer do you come up with? There are probably a number of reasons why you are here. Some of the reasons participants give for involving themselves in a human-relations-training laboratory are

- My friends are doing it.
- I want to find out how others see me.

- It's a required course.
- I want to check out the quality of my interpersonal skills.
- Others have told me that it's a good course.
- I feel inadequate in interpersonal situations.
- I'm curious to see what such a laboratory experience is like.
- I have a general desire to grow; this course sounds like it might help.
- I want to develop some basic interpersonal skills.
- I'm lonely; I feel interpersonally alienated.
- I want to become more assertive in interpersonal situations.
- I'm interested in examining my interpersonal style.

This laboratory is based on the contract outlined in Chapter 2. You may be more or less ready to give yourself to the laboratory experience in the ways outlined in the contract, but some minimal acceptance of the provisions of the contract is essential. No one can provide you with the motivation you need to make the laboratory experience a growthful one. That must come from you.

SELF-CONCEPT AND INVOLVEMENT WITH OTHERS

How I see and feel about myself affects how I involve myself with others. Consider a few examples.

- If I feel that I am "no good," I will hesitate to seek out others as friends, for friendship is a kind of "gift of self" (and who would give someone a defective gift?). I will also tend (subtly or otherwise) to repulse overtures to friendship made by others, because I feel "unworthy."
- If I see myself as a growing person (a person with B-aspirations, and not just someone mired down in D- or M-concerns), I will take initiative in interpersonal transactions.
- If I see myself as weak and needy, I will probably present myself to others as a dependent person.
- If I see myself as riddled with problems with which I cope poorly, I will present myself to others as a "helpee" and will expect others to become not my friends but my helpers.
- If I feel good about whatever strengths and resources I have (for instance, I see myself as a caring person) and do not need to assure myself by constantly comparing myself with others, then I am free both to accept others and to challenge them to grow.
- If I am a male and feel insecure about my masculinity, then I may take a Don Juan stance toward women, trying to manipulate them as objects rather than involving myself with them as persons.

During the laboratory, then, it is important for you to examine your own feelings about yourself as you examine your interpersonal style, for your self-concept greatly influences your style. The feedback you receive from others as they observe you interacting (or failing to interact) in the group will help you both confirm what you discover about your self-concept and become aware of possible blind spots. You might hear such things as

- The way you apologize for yourself makes me wonder whether you really like or respect yourself.
- I notice that you single out the more attractive people in the group to interact with. This makes me wonder whether, at least unconsciously, you consider yourself superior or better than others in some ways.
- You take risks in revealing yourself in the group without in any way being overly dramatic or exhibitionistic. This makes me think that you feel at home with yourself, that you basically accept yourself even though you know there's room for improvement.

Self-esteem is one of the most important dimensions of your self-concept. Your ability both to develop interpersonal skills and to use those you do develop depends, to a great extent, on the quality of your self-esteem. Brandon (1971) claims that self-esteem is the most important single key to behavior. Simpson and Hastings (1974) define it as

> the evaluation that an individual makes and customarily maintains in regard to himself. It is a personal judgment of his worthiness as a person, indicating the extent to which he believes himself to be capable, significant, and successful. Beyond this general definition, self-esteem can be divided into two components: a sense of efficacy or self-confidence based on feelings of competence and a sense of worthiness or self-respect based on feelings of worth [Simpson & Hastings, 1974, p. 174].

When you find yourself having difficulty engaging in the kinds of interpersonal behavior suggested in the chapters of this book, try to find out whether your feelings are getting in the way. If you see yourself as a mousy, impotent person, it only stands to reason that you will find it very difficult, if not impossible, to challenge others, even when others let you know that they want to be challenged. It is not enough to see yourself as others see you. It is necessary to know how you experience yourself and the ways in which this self-experiencing enhances or detracts from your interpersonal involvement.

A SKILLS APPROACH

The subtitle of this book includes the word "skills." According to Maslow (1968, 1970), in order to grow you must satisfy a variety of human needs. Interpersonal skills can be seen, from one point of view, as instruments enabling you to satisfy a number of your human needs. If you have the skills to express yourself, to respond to others, to place legitimate demands on others, and to open yourself up to being influenced by others, these skills help you to satisfy

- *safety needs:* the need for a world that is secure, orderly, dependable, and free from threat; and
- *esteem needs:* the need for both self-respect (for instance, the kind of self-respect that comes from a sense of interpersonal competence) and for respect from others.

You are a social being with social needs that are fulfilled through interaction with others. Successful interaction requires the development of effective interpersonal skills. Therefore, interpersonal skills are important not only to help you gratify legitimate personal needs but also

- to enable you to gratify the legitimate needs of others;
- to help you prevent or work through interpersonal conflict;
- to enable you to establish effective working relationships; and
- to help you develop feelings of interpersonal competence [see Simpson & Hastings, 1974, p. 64].

It is not suggested here that interpersonal skills can be separated from the rest of your life. Interpersonal skills are used in the context of the whole person. Therefore, your interpersonal skills are influenced by such factors as

- *motivations:* you may know how to confront a friend in a caring, responsible way, but you may choose to confront him in a punishing way.
- *needs:* you may know how to disclose yourself appropriately to others, but you may fear to do so. You have grown up with certain security needs unfulfilled.
- *personality characteristics:* you may know how to express warmth and affection, but you seldom do so, because you have developed the interpersonal style of a "withholder" rather than a "giver."
- *attitudes:* you may know how to confront well, but you do not, for you see confrontation as a "lack of charity" toward another person.
- *values:* you may know how to communicate empathic understanding for others, but you seldom do so, because providing support consistently to others is not a behavior you prize.

Thus a human-relations-training laboratory experience that emphasizes skills means that you will have an opportunity to take a look at your level of competence in a variety of basic interpersonal skills, that you will be trained in the skills you lack or in which you are weak, and that you will be given an opportunity to use these skills in your interactions in the group.

INTERPERSONAL STYLE

Since the term "interpersonal style" occurs frequently in this book and is central to the definition of the goals of the laboratory in human relations described here, it is a term that must come alive for you. Interpersonal style may be defined as

> your characteristic ways of thinking and feeling about and interacting with other people together with your interpersonal skills (and lack of skills) and your characteristic successes and failures in your human relationships.

But this definition is still not concrete and usable enough. Since interpersonal relationships are complex, it is very difficult to present a neat categorization of the dimensions of interpersonal style. However, you can make the concept of your own interpersonal style more concrete for yourself by considering the categories below and by asking yourself the questions within these categories. You will soon discover that the categories listed are not exhaustive (there are many ways in which you can look at your interpersonal style) and that they overlap (a sign of the complexity of human interrelating). Still, if you do ask yourself these questions, you will get a much more concrete feeling for the term "interpersonal style" and, better yet, you will begin to get some feeling for your own interpersonal style. Since the contract in Chapter 2 asks you to discuss your interpersonal style concretely with your fellow group members, answering these questions will prepare you for group interaction.

Extensiveness. How extensive is my interpersonal life? How much of my day is spent with people? Do I seek out opportunities for being with people? Do I have many friends and acquaintances or few? Are my contacts with others planned or left up to chance? Is my life too crowded with people? Are there too few people in my life? Do I prefer smaller gatherings or larger groups? Do I have a need for quiet time away from people? Am I outgoing or introverted or somewhere in between?

Needs and wants. What are my interpersonal wants and needs? How do I express them? Directly? Indirectly? Do I like to be challenged? Complimented? Reassured? Left alone? Treated like a child? Like a parent? Do I like to be responsible and assertive? Do I want others to control me? With what kind of people do I associate? Are they like me? By what criteria do I choose my friends and acquaintances: chance, intelligence, physical attractiveness, good-naturedness, values, social position?

Caring. How caring am I in my interactions with others? Do others know that I care? Is it obvious in my behavior? Do I sometimes wonder whether I care at all? Do I take others for granted? Who really cares for me?

How do I show care? How is care shown to me? In what ways am I self-centered? Am I a generous person? How do I express my generosity?

Competence. What are my interpersonal skills? Am I good at communicating understanding to others? Am I appropriately warm? Do I communicate to others that I respect them? Am I my real self when I am with others—that is, do I communicate genuineness? Am I open, willing to talk about myself appropriately? Can I challenge others and invite them to explore their behavior without being accusatory or punitive? How effective am I in exploring my relationships directly with others? What skills do I want to acquire? How well do I meet strangers? Am I awkward, embarrassed, resentful? Or enthusiastic, poised, confident?

Emotions. What do I do with my emotions in interpersonal situations? Do I swallow them? Or some of them? Do I wear my emotions on my sleeve? Is it easy for others to judge what I am feeling? Am I moody? To what extent am I ruled by my emotions? How do I make my emotions public? How do I feel about being emotional in interpersonal situations? How perceptive am I of the emotional states of others? How do I react to others when they are emotional? What emotions do I enjoy in others? What ones do I fear? What do I do when others keep their emotions to themselves?

Intimacy. How intensive is my interpersonal life? Do I actively pursue intimacy with others? Do I encourage others to get close to me? If so, how? Are there people with whom I would like to be more intimate? Are there people who will not allow me to be intimate with them? What does intimacy mean for me? Do I see that there are a variety of ways of being intimate with others? What kind of people are attracted to me? What forms of intimacy do I find most rewarding? Most threatening?

Rejection and alienation. Is loneliness ever a problem for me? If so, how do I handle it? Do others see me as lonely at times? If so, how do they react to me? What do I do when I see that others are lonely? Am I easily threatened by others? How do I react when I am threatened? What threatens me in interpersonal situations? Have I experienced rejection? How do I handle being ignored or left out or rejected? Do I ignore or reject others? How do I handle the problem of not wanting to relate to people who want to relate to me?

Interpersonal influence. What demands do I place on my friends and/or acquaintances? Am I manipulative in my dealings with others? If so, with whom and how? Do others try to manipulate me? Who and how? What demands do my friends place on me? Do I tell others explicitly what I expect from them, or do I assume that they know what I want? Do I see that giving and receiving influence can under certain conditions be proper and growthful in my interpersonal life? If so, under what conditions? Am I either dominant or submissive in my relationships to others? When and under what conditions?

Mutuality. Do I allow for give and take in my relationships with others? Am I authoritarian or parental? Am I democratic? Am I laissez-faire? Am I willing to compromise? Do I take responsibility for what happens in my relationships, or do I "let things take their course"? Do I encourage mutuality in decision making? Do I encourage dialogue? Do I expect to be treated as an equal? Do my relationships involve mutual responsibilities?

Work relationships. How do I get along in my work relationships? Do I treat people at work as people or as roles? Do I assert myself with my superiors (including teachers)? Am I understanding with my subordinates? Do I have prejudices toward people in certain "inferior" roles? Am I overpersonal at work? Am I a loner at work?

Values. What are my principle interpersonal values? Caring? Self-interest? Solid work relationships? How open to interpersonal growth am I? What am I willing to risk with others? In what areas am I not willing to take risks? Do I allow ambiguity and uncertainty in my interpersonal life? Am I tolerant of others whose opinions differ from mine? What are my prejudices? Am I willing to change my own values, beliefs, and behaviors when, in my dealings with others, it seems appropriate to do so? How rigid or flexible am I in my relationships with others? Do I seek out ways to grow with others? Do I share my values with others? Can I put my interpersonal relationships into perspective by putting them into the wider contexts of work, world conditions, and so on?

These questions may leave you bewildered because there are too many of them, because you have not reflected much on them, or because they are not the right questions for you. However, they do attempt to get at your characteristic ways of thinking and feeling about and interacting with other people; that is, they try to get at the dimensions of your interpersonal style. It is hardly necessary for you to answer all of these questions right now, but it is important to know what is meant by "interpersonal style."

EXERCISES

Below are the first of many exercises. These exercises are designed to help you achieve the behavioral goals outlined in each chapter. For instance, the exercises below are designed to help you get a feeling for the term "interpersonal style" and, more concretely, a feeling for some of the dimensions of your own interpersonal style. Some of the exercises are designed to stimulate your thinking about yourself, your interpersonal style, and your relationships to your fellow group members. As such, they help you to come to the group with an agenda—that is, something to say about yourself and something to say to your fellow participants. Other exercises are designed to help you practice, on paper, the core interpersonal skills discussed and illustrated in this book. Ordinarily, the instructor will indicate which exercises are to be done. The exercises are suggested as ways of making the theory of this book more concrete for yourself. They should be used to the degree that they fulfill this function.

Exercise 1: Your Own Interpersonal Style

a. Pick out the questions in the list above that you would like to answer for yourself in more detail. Share these with your fellow group members without making any extensive attempt now to answer them. This exercise will give you and your fellow participants some idea of how many different individual "agendas" exist within the group.

b. Make a list of questions or categories, or both, that you think are missing in the listings above. Try to think of questions that, if answered, would help you get a better perspective on your own interpersonal style. Share these new categories and/or questions with your fellow group members.

Exercise 2: This Is Me

Using the questions above and your own set of questions as a guide, write a preliminary one-page description of your present interpersonal style. Share this description with the members of your group.

CHAPTER 1: FURTHER READINGS

On the laboratory approach to learning:

Benne, K. D., Bradford, L. P., Gibb, J. R., & Lippitt, R. O. (Eds.). *The laboratory method of changing and learning: Theory and application.* Palo Alto, Calif.: Science and Behavior Books, 1975.

Bradford, L. P., Gibb, J. R., & Benne, K. D. (Eds.). *T-Group theory and laboratory method.* New York: Wiley, 1964.

Egan, G. Introduction; The laboratory method. Chapters 1 and 4 in *Encounter: Group processes for interpersonal growth.* Monterey, Calif.: Brooks/Cole, 1970. Pp. 1–24; 104–122.

Golembiewski, R. T., & Blumberg, A. *Sensitivity training and the laboratory approach: Readings about concepts and applications* (2nd ed.). Itasca, Ill.: Peacock, 1973.

Howard, J. *Please touch: A guided tour of the human potential movement.* New York: McGraw-Hill, 1970.

Rogers, C. *On encounter groups.* New York: Harper & Row, 1970.

Shaffer, J. P. B., & Galinsky, M. D. T-Groups and the laboratory method. In *Models of group therapy and sensitivity training.* Englewood Cliffs, N. J.: Prentice-Hall, 1974. Pp. 189–210.

On caring as a value in interpersonal living:

Mayeroff, M. *On caring.* New York: Harper & Row, 1971.

On values and their clarification:

Raths, L., & Simon, S. B. *Values and teaching.* Columbus, Ohio: Charles E. Merrill, 1966.

Rokeach, M. *The nature of human values.* New York: The Free Press, 1973.

Simon, S. B. *Meeting yourself halfway: Thirty-one value clarification strategies for daily living.* Niles, Ill.: Argus Communications, 1974.

Simon, S. B., Howe, L. W., & Kirschenbaum, H. *Values clarification: A handbook of practical strategies for teachers and students.* New York: Hart, 1972.

On self-actualization:

Maslow, A. *Toward a psychology of being* (2nd ed.). New York: Van Nostrand-Reinhold, 1968.

Maslow, A. *Motivation and personality.* New York: Harper & Row, 1970.

On self-concept and self-esteem:

Brandon, N. *The psychology of self-esteem.* New York: Bantam Books, 1969.

Brandon, N. *The disowned self.* New York: Bantam Books, 1971.

Simpson, C. K., & Hastings, W. J. Love thyself: Esteem workbook. Chapter 10 in *The castle of you: A personal growth workbook.* Dubuque, Iowa: Kendall/Hunt, 1974. Pp. 173–189.

Chapter 2

The Contract: An Overview of the Human-Relations-Training Model

This chapter contains the contract that provides the structure for the human-relations-training experience you are entering. You have already read a short discussion on the use of structure in human-relations-training groups. It is hoped that the structure presented here through the provisions of the laboratory contract will not tie up your energies. Spontaneity without discipline, however, is irresponsible. The disciplined person masters his life and has, therefore, the time and the energy needed for spontaneity. The undisciplined person may seem to be living spontaneously, but actually he or she is living haphazardly; and this way of living is only a caricature of spontaneity. The contract demands discipline on your part, not for its own sake but in the service of learning interpersonal skills. Someone has said that no one can confer freedom on another person; freedom is something that each person must work for, must seize for himself. In this, interpersonal competence is like freedom. It cannot be conferred; it must be worked for. The structure presented by the contract is designed to help you channel your energies as you work toward greater interpersonal competence.

THE CONTRACT

The Goals of the Training Group

It has been pointed out that, in both ordinary day-to-day interpersonal transactions (Shapiro, 1968) and in interpersonal-growth-oriented laboratory experience (Egan, 1971a), unwritten and unspoken contracts regulate and govern much of what happens. The trouble with these ever-present unwritten and unspoken contracts is that they introduce inequities into the social situation, causing people to run afoul of one another. Let's consider a couple of examples.

- *Spouse A* (Unconsciously, preconsciously, or subconsciously to self): In our marriage I will relate primarily to you, but I also maintain the option of establishing close relationships with others of both sexes.
- *Spouse B* (also to self): Ours will be the great friendship. Others will be acquaintances, fairly peripheral to our lives.

- *Lab participant A* (to self): I will play it cool, not involve myself very much, until I see what's going to happen here.
- *Lab participant B* (to self): The only way to survive a laboratory experience is to be utterly, even brutally, frank and honest. I'll tell everybody how I feel about anything that happens.

Obviously, such unformulated and unspoken unilateral interpersonal and social contracts can only cause trouble. Therefore, in order to ensure that you and your fellow participants are "buying into" the same kind of experience, the goals of this laboratory are presented in the form of a social contract. This contract does not demand that you, as an individual participant, surrender your identity, or that you give up the right to pursue your individual goals. It does demand that you make your individual goals clear to the other participants, and that you pursue them within the framework of the general social contract.

As you enter this group experience, you contract for a twofold goal.

1. Exploration. You will use your time in the group, first of all, to examine your own interpersonal style (in the concrete, operational terms described in Chapter 1). This process is designed to help you come to a practical awareness of your characteristic ways of relating to others, including both your strengths and your weaknesses.

2. Experimentation. You will also use your time to alter your interpersonal style in ways you deem appropriate. One mode of altering will be to check out and strengthen basic interpersonal skills. Your work in the group should help you to consolidate and develop your interpersonal strengths while beginning to work at eliminating or coming to grips with your weaknesses.

How Are These Goals to Be Achieved?

You contract not only for the goals outlined above but also for a process to help you achieve these goals. These are the provisions of the contract.

1. Establishing and developing relationships. Your first—and overriding—means of achieving these goals is to participate actively in the process of establishing and developing relationships with your fellow group members. This process demands that

a. In everything you do, throughout the training and in all of the exercises, you are attempting to establish and develop a relationship with every other member of your group.
b. As you move through the process of attempting to establish and develop these relationships, you observe at first hand your own interpersonal style.
c. At the same time, you receive feedback from your fellow group members on your style, including your strengths and your weaknesses.
d. You have the opportunity to experiment with "new" behavior—that is, to attempt to alter dimensions of your interpersonal style in order to become, in your own eyes, more interpersonally effective.

2. The core interpersonal skills. You will be trained in a core set of basic interpersonal skills. This preliminary set of skills will include

a. *Self-presentation skills.* Included here are the skills of appropriate self-disclosure, concreteness, and expression of feeling.
b. *Response skills.* Included here are the skills of attending and listening, the communication of empathic understanding, and the behavioral communication of genuineness and respect. Training in these two groups of skills will take place through the subgroupings made up from your basic training group.
c. *Challenge skills.* You will be trained in a set of advanced skills, including skills of higher-level empathic understanding, confrontation, and immediacy (direct "you-me" talk). Much of this training will take place in the larger group.
d. *Group-specific skills.* You will be trained in how to use both self-presentation and response skills in the larger group. Initiating is more difficult in a group than in one-to-one dialogue.

Developing Relationships

The phrase "to establish and develop relationships of some closeness" will be used repeatedly in these pages. Let's get a more concrete idea of just what this means (after all, it is one of the major goals of the laboratory). In everyday language, establishing a relationship means

- spending some time with another
- doing things with another (talking, sharing)
- beginning to feel comfortable in talking about relatively important issues
- developing a respect and perhaps a liking for another
- caring about, being concerned about another
- developing a sense of give-and-take, or mutuality, in sharing
- being willing to discuss at least certain problems with another

- feeling comfortable with another
- being willing to help another (even though you are not there primarily as "helper")
- possibly doing things outside the group with another

Write down meanings you feel should be added to this list. Look at the relationships you have already established in your life and try to write down, in behavioral terms, what it has meant to establish and develop a relationship. Share your list with one or more members of your group.

Obviously, you will not establish and develop relationships in this group in exactly the same way you do in your day-to-day life. First of all, in day-to-day life establishing relationships is often something that just "happens." However, even in "real" life there may be times when you set out to establish a relationship with someone. You do so because you are, ordinarily, attracted to the other person, and therefore you want to establish deeper contact. In this group, however, you are asked to establish and develop relationships just because these other people are members of your group and, like you, interested in improving the ways in which they relate to others. The time you will spend with these people in any given week is limited, and most of these relationships will most likely end (at the end of the semester)—although in a significant number of cases more enduring relationships are established. Therefore, there is something quite pragmatic about these relationships; you are establishing and developing them for a particular purpose. This does not mean that what you are doing is phony, even though it is somewhat artificial and different from what you do "naturally" or ordinarily.

Let's review once more what it means to "establish and develop a relationship" from the viewpoint of *investing* in others. You are asked to invest yourself in your fellow group members, at least for the given period of time, in ways similiar to the ways you invest yourself in your friends. This investment means

- working cooperatively with others in trying to achieve the goals of the group;
- learning how to listen to others, to be actively with others as they talk about themselves;
- responding to others concretely, about both the feelings and the content (behavior, experience) that constitute their messages;
- gradually letting others into your "world"—that is, talking about yourself in ways that help promote the goals of the group (such as talking about your interpersonal style as it now stands and how you feel about it);
- letting others know what you like about them, what you see them doing well in relating to you and to others;
- letting others know what holds you back from getting involved with them—what they do that scares you, annoys you, causes you to withdraw;

- sharing what you learn about yourself during the course of the laboratory;
- letting others know how you would like to change certain behaviors in your interpersonal style—what you would like to drop and what you would like to add;
- letting others help you place these demands for change on yourself;
- asking for feedback on your own interpersonal style and the quality of your interactions in the group; and
- dealing with feelings that arise from participation in the group and that affect the quality of your participation in the group.

Perhaps you can think of further ways of investing yourself in others and having them invest themselves in you. The skills that will be discussed in many of the pages of this book can be seen as ways of your investing mutually in one another. It should be evident by now that establishing and developing relationships means that you are to be *proactive* (initiating), and not just reactive, in the group. The assumption here is that establishing and developing relationships is a great deal of work. Indeed, laziness is one of the principal enemies of more intensive interpersonal relating; so often one's comfort seems more important than one's relating to others. A sterile interpersonal life is the price of excessive interpersonal comfort. It is not suggested that you will come to like everyone in your group, in the sense that you would like to continue to develop your relationship outside the group. However, you should come to *respect* the other members of your group, and one way of showing respect is by working with them to achieve the goals of the group.

A THREE-PHASE MODEL

This training program has three phases or stages.

Phase I, Part I: Learning the Skills of Relationship-Building: Trust and Risk

Much of the training in this phase takes place in one-to-one conversations with an observer present. Since relationship-building involves both trust-generating mutual understanding and the risk of self-disclosure, you will be trained in both sets of skills.

A. The Skills of Self-Presentation

1. Self-disclosure. Through a variety of exercises, you will be trained in appropriate self-disclosure (as opposed to "secret-dropping" or exhibitionism or unproductively keeping things to yourself). Since your goal is to examine your interpersonal style, your disclosure should be related to, and should help you achieve, this goal.

2. Concreteness. You will learn what concreteness in interpersonal dialogue means, and you will have the opportunity to practice being concrete about your own experiences, feelings, and behaviors (as related to appropriate self-disclosure). The more concrete your statements are, the more immediate your interactions with others become.

3. The expression of feeling. You will learn how to identify and express your feelings more carefully. This does not mean that you will be asked to "perform" by engaging in emotional displays. Rather, you will begin to discover the place of emotion in your interpersonal dialogues.

B. The Skills of Responding

In Phase I you will also practice the skills that are correlated to the skills of self-presentation—that is, the skills of responding. When you reveal yourself, you expect to be understood. Therefore, you should also learn to actively exercise response skills.

1. Accurate empathy. You will be trained in the skill of accurate empathy—that is, the ability to respond actively and with understanding to those who disclose themselves. Accurate empathy is perhaps the most critical of all interpersonal skills.

2. Respect. The lab is an opportunity to examine the quality of your respect for others. Failure to respect others will show up in your behavior. If you want to respect others, you will learn how respect can be communicated in concrete, behavioral ways. Respect as a moral quality remains lifeless unless it is communicated behaviorally to others.

Although Phase I is an individual skills-training phase, the goal of establishing and developing mutual relationships is still in effect. The skills of Phase I are the core skills of relationship-building. In the practice sessions you do not merely role-play, for the content of your self-disclosure (dealing with your interpersonal style) is real, and your response to your fellow group members is also real.

Phase I, Part II: The Skills of Challenge

Once you come to realize the importance of relationship-building skills and begin to become adept in them, you can turn your attention to the skills of challenge. The best relationships include challenge as well as understanding, support, and fellowship. Those who care about you most will also challenge you to use your resources to live life more fully.

Part II of Phase I helps you learn how to challenge others responsibly. Part I of Phase I helps you explore your interpersonal style from your own (inner) frame of reference. You begin to see more clearly how you act in interpersonal situations and how you feel about your interpersonal transactions. However, there are frames of reference other than your own. As others watch your style, they are in a position to give you feedback, to challenge you to explore your interpersonal behavior more carefully, and to help you place on yourself the demands for interpersonal growth to which you yourself are committed. New skills are needed in order to engage in this process of feedback and challenge responsibly and growthfully. Let's examine these skills.

1. Identifying strengths. As you observe your fellow participants interact, you begin to see not only what they fail to do or do poorly, but also what they do well—their interpersonal strengths. It is a skill to be able to identify the strengths of others and to let them know the ways you see these strengths operating. In fact, if you cannot appreciate the strengths of others, it is probably better not to say anything about their weaknesses. However, if you can let others know that you appreciate their strengths, they will be much more open to hearing something about their weaknesses from you.

2. Advanced accurate empathy (AE II). By means of this skill, you communicate to the other person an understanding of not only what he actually says but also what he *implies*, what he hints at, what he fears to state more clearly. You help the other person (and he helps you) begin to make connections between seemingly isolated statements, and these connections lead to deeper self-understanding.

3. Confrontation. You learn how to invite others to look at their behavior more carefully, especially in terms of discrepancies (what a person says versus what he actually does, for instance) and unused interpersonal strengths and resources.

4. Immediacy. Immediacy is direct "you-me" talk. It is the skill of being able to examine with another person what is happening in the here and now of your relationship with him. It is a complex skill, for it involves, ordinarily, the ability to combine advanced accurate empathy (your vision of what is happening in the relationship), self-disclosure (how you feel about the relationship), and confrontation (inviting the other to explore the potentialities and problems of the relationship).

These skills constitute a kind of strong medicine in human relationships and have to be used carefully. Therefore, in Phase II, you are provided not only opportunities to learn and exercise these skills but also opportunities to *plan* their use and to get feedback on how successfully you use these skills.

Phase II: Group-Specific Skills

Neither the skills of relationship-building nor the skills of challenge, if learned in one-to-one interactions, necessarily generalize to a group situation. Therefore, all skills need to be practiced in the group itself. Learning how to use a variety of skills in a variety of combinations actively in the group is what is meant by the term "group-specific skills." Let's examine some of these skills.

1. Responding actively. Responding actively means that when you are contacted directly by someone else in the group you respond actively—that is, contribute actively to the dialogue. You are not in the group to be dealt with (this role is too passive and makes you a "patient"). You are there to deal with yourself and your interpersonal style actively. Therefore, it does not make any difference who initiates a dialogue. When dialogue takes place, both participants are actively responsible for it.

2. Taking initiative in the group. Since you have the task of trying to establish and develop a relationship with each of the other members of your group, it is not enough that you merely respond (even actively) when others contact you. It is also essential that you take the initiative to contact others without being contacted first. It takes practice to feel at home with venturing out and initiating dialogue with another. What are these initiating or "venturing" skills?

a. Using primary-level accurate empathy. When others talk, take pains to both listen and respond with accurate empathic understanding (AE I). In Phase I you were "forced" to do this, for at times you were in the role of "respondent" in the practice sessions. Now you must initiate this kind of response on your own. AE I is extremely important throughout the life of the group, for it maintains an atmosphere of trust and support. Individual members must supply it spontaneously.

b. Self-disclosure. In Phase I you had to assume the role of "speaker" at times and thus to disclose yourself. In Phase II, you must reveal yourself appropriately without being requested to do so. Each group member has to ask himself continually: "What do I want to reveal about myself that will help me and others achieve the goals of this group?"

c. Owning the interactions of others. When two or more of the other members of the group are engaged in conversation, it is not enough just to listen well. Rather, when it is appropriate, you must take the initiative to intervene ("own" their conversation), not by disrupting their dialogue with your own agenda, but by helping them to get where they are trying to go in their dialogue—helping, that is, by understanding (AE I), sharing your own experience, giving your own reactions, offering new perspectives, confronting, and so on. If you do this skillfully, you will contribute and not just interrupt.

d. Using challenging skills. In Phase II you use the skills of challenge actively, with discretion, to help others see themselves from frames of reference other than their own. Here the skill of immediacy becomes especially important, for through it you check to see what is happening in each of the relationships you are trying to develop ("Where do you and I stand now in our relationship?").

e. Calling for feedback. Instead of waiting to get feedback from others, you can call on others to be your "consultants." That is, you can ask others for confrontation, immediacy, and any other kind of interaction you feel would help you achieve your goals.

In Phase II you can expect a certain degree of "interaction self-consciousness"; that is, you will still be aware of the fact that you are to one degree or another practicing skills. And, although there is less structure in Phase II than in Phase I, a degree of structure remains that adds to this kind of self-consciousness.

Phase III: Pursuit of the Core Contract

The core contract includes

- examination of your interpersonal style
- by both observing yourself and getting feedback, as you establish and develop relationships with your fellow group members;
- acquiring and/or strengthening your basic interpersonal skills; and
- beginning to alter your interpersonal style in ways that you see would make your interpersonal life more effective.

In Phase III, structure is reduced to a minimum, and the members of the group pursue the core contract through an "open" group experience, using the skills they have acquired or are acquiring.

It is hoped that by Phase III you will have acquired enough initiative or agency to become an independent and active contributor to the group experience. Although some focused exercises or structured experiences may be used during this phase, the emphasis is on the initiative and skills of individual group members.

Phase III gives you an opportunity to take a closer look at the problems you experience in interpersonal living, as they are manifested both in the training group and in your day-to-day life. Not only problems, but also action strategies to meet these problems, can be considered. Skills training is in itself a very potent action strategy, one that has immediate transfer value outside the group. People with problems in interpersonal living often find that they begin to transcend many of these problems once they have the tools of effective communication. However, Phase III does not limit itself to a consideration of problems. Many of us are too problem-oriented and too often end up as "helpers" or "helpees" in our dialogues with others. The lab gives you an opportunity to explore self-actualizing goals (Maslow, 1968) in the area of interpersonal living. How intensively do you want to live interpersonally? How do you live a more intensive interpersonal life without taking interpersonal living out of the context of the rest of life? Phase III is also a good time to determine what unused interpersonal resources you have. You can begin to develop them and bring them to bear, first on your interactions in the group itself, and then on your day-to-day interactions as well.

In a group of eight members, there are 56 different sets of relationships to be worked through (or 28 separate "pairings"). This means that in the average group there is a great deal of work to be done if the core contract is to be fulfilled—perhaps more work than can be accomplished in the ordinary time-limited group experience. Therefore, even in a highly active group, there will be unfinished business at the end.

You will probably find that some skills come more easily for you than others, and it will take work to achieve a well-rounded skills repertoire. You will also probably develop better relationships with some members than with others. However, there is much to be learned equally from successes, partial successes, and even failures in the process of establishing and developing relationships.

Exercise 3: A Log

Keep a log of the thoughts, feelings, experiences, and behaviors that highlight each meeting and of the thoughts and feelings you have about the group between sessions.

Enter material you can use to make the next meeting a more effective here-and-now learning experience for yourself and your fellow group members. Enter *experiences* ("Jane ignored me the whole meeting. In general she has shown a certain indifference toward me. Check to see what is going on"), *behaviors* ("I asked John a lot of questions and really did not make much of an effort to understand him. I noticed during the week that I do that quite a bit. I think others should challenge me more when I act like that"), and *feelings* ("I've been on a 'high' from the last meeting; everyone in the group contacted me, but no one dealt with me as if I were a 'case,' even though I cried. I don't want to be a blubbering slob, but I want to be able to cry at times without feeling I'm betraying my manhood").

Keep track of what you have to work on and put effort into it (for example, using accurate empathy more frequently, not avoiding people who seem distant to you, and so on).

Use the log to keep track of where you stand with each of the other members in terms of establishing and developing relationships.

Make your entries relatively brief and concrete. Ask yourself whether you can use what you write at the next meeting.

There is a tendency on the part of some participants to keep excellent logs but then to fail to use this material in the group meetings. If you are having difficulty using your log material, perhaps it is good to make this problem known at a meeting and let others help you introduce the material into the group discussion.

In your log keep a separate page for each of your fellow group members. Keep all of your reflections on each person on his particular page. This will give you a cumulative record of the highlights of your interactions with each. Glancing at each page, you will be able to see just how your relationship has developed (or failed to develop) with each. Not only will this exercise enable you to learn about your interpersonal style, but it will also help you greatly in giving feedback to your fellow group members.

Draw an agenda from your log. Your log has a very practical function in relation to the group. As you read your log, you can come to some decisions on what you want to accomplish *in the next group meeting.* Therefore, each weekly log should conclude with a *practical* agenda for the next group meeting. For instance, you might write in your log:

> I don't talk to Jane at all, because I think she is rather indifferent to me and I'm attracted to her. I don't like this combination.

Then your agenda at the end might have the following entry:

> Talk to Jane. Tell her your feelings. Clear the air. It's no use to merely avoid her, and you must admit you don't really know how she feels.

Concrete logs lead to concrete agendas. I don't mean to imply here that it is easy to put concrete agendas into practice in the group. I suggest, however, that practical, concrete agendas increase the probability of your involving yourself more directly and concretely with your fellow participants.

The log together with an agenda for the next meeting is, then, not a one-time exercise. It is a continuing exercise and perhaps one of the most important ones you will do. In unstructured groups the members usually come unprepared to group meetings. Each member could probably say to himself or herself: "I wonder what we're going to do in this meeting." The log/agenda exercise will help you *make* things happen during your group sessions instead of just *allowing* things to happen. It will reduce the amount of time that you and your fellow group members mill around and waste time.

CHAPTER 2: FURTHER READINGS

On contracts in groups and in interpersonal living:

Carney, C., & McMahon, S. L. The interpersonal contract. In J. W. Pfeiffer & J. E. Jones (Eds.), *The 1974 annual handbook for group facilitators.* La Jolla, Calif.: University Associates, 1974. Pp. 135–138.

Egan, G. The contract. Chapter 2 in *Encounter: Group processes for interpersonal growth.* Monterey, Calif.: Brooks/Cole, 1970. Pp. 25–67.

Egan, G. Contractual approaches to the modification of behavior in encounter groups. In W. A. Hunt (Ed.), *Human behavior and its control.* Cambridge, Mass.: Schenkman, 1971. Pp. 106–127.

Shapiro, S. B. Some aspects of a theory of interpersonal contracts. *Psychological Reports,* 1968, *22,* 171–183.

PART 2

Phase I: The Skills of Letting Yourself Be Known

A GENERAL INTRODUCTION TO TRAINING IN INTERPERSONAL SKILLS

Participation in a small-group experience that focuses on one's current interpersonal style and the possible modification of some dimensions of that style is a demanding, high-level communication experience. Such an experience requires a definite set of skills on the part of the group members if it is to be done well. My experience shows me that to assume that all of the members of any given group possess these skills from the beginning of the laboratory is unwarranted. Even when all of the members of the group agreed beforehand to live up to a certain "contract" (Egan, 1970, 1973a), many of them failed to do so. My first instinct was to suspect ill will, but eventually I realized that the real problem was lack of skill. People were being asked to do something (fulfill the contract) for which they were not prepared. For this reason, I expanded the human-relations-training model into a two-phase approach (Egan, 1973b) consisting of an individual skills-training phase plus an open-group contract phase. (A further problem still remained, but this problem will be considered in Phase II.)

In Phase I, no assumptions are made with respect to the quality of your interpersonal skills; that is, you are not required to "perform" at any particular level. This aspect of the lab structure is in keeping with the noncompetitive nature of the group experience. All members of the group, whatever their level of competence, go through the skills-training phase of the program. This does not mean, however, that members with relatively highly developed skills are merely treading water, for the goals of Phase I are multiple, as described in the paragraphs that follow.

Review of skills competence. Practice in the two sets of basic relationship-building skills provides you an opportunity to check out your level of competence in these skills. It is a kind of interpersonal-skills checkup analogous to a physical checkup. This goal is common to all participants, no matter what their level of skills competence.

Acquisition and strengthening of skills. If you have some of the Phase-I skills, then you can use this time to strengthen them. If, on the other hand, you discover that you either lack a specific skill or are deficient in the exercise of this skill, you can use the time to remedy the deficiency. This goal, too, is usually common to all participants, but to different degrees. No one has it "made" in interpersonal skills. For instance, you may discover that you can use your skills quite effectively with one kind of person (for instance, a person who is somewhat dependent) but have a much more difficult time with another (an aggressive person).

Experience of a training methodology. Everyone in the group is learning a skills-training methodology by actually experiencing it. Your experience will eventually put you in the position of helping others acquire these skills.

Self-exploration. Since you will be talking about various aspects of your interpersonal style, the skills-review/skills-acquisition phase will also provide you with an opportunity to explore your style from your own frame of reference. Since this phase emphasizes respect, support, nonjudgmentalness, and understanding, you will have the safety you need to explore your style.

Relationship-building. Phase I is not just practice. Although it is relatively heavy on structure, this structure gives impetus to the process of establishing and developing relationships among group members. If your skills are good to begin with, you can use them to understand your fellow participants and to help them understand themselves.

Helping others to develop skills. Finally, if you are highly skills-competent at the beginning of the laboratory, you can help your fellow group members acquire and develop their skills. If you become a nonpatronizing helper (or ancillary trainer) in the skills areas in which you are competent, you will help establish a climate of cooperation instead of competition. You will contribute to the group as a member of a learning community.

The Effective Use of Interpersonal Skills

There are three components or dimensions to the successful exercise of interpersonal skills. Since each component is necessary, care must be taken during training to develop each dimension.

1. *The awareness or discrimination dimension.* One of the basic factors in social intelligence is the ability to see what is called for in any given interpersonal or social situation. Social intelligence refers, for instance, to your ability to see what emotional struggle a person is going through and to understand the experiences and/or behaviors from which his emotions stem. Or it may refer to your ability to see whether support and understanding or confrontation and challenge would be more useful in the here and now of

your interaction with another person. This kind of discrimination or awareness will be discussed at greater length when "attending" is considered.

2. *The technology dimension.* Once you perceive what kind of interaction is called for in a given interpersonal situation, you must have the "know-how" to deliver what is called for; that is, you must be technically competent in the skill. For instance, if accurate empathy is called for, you must be able, first, to correctly identify the speaker's emotions and the experiences and/or behaviors underlying his emotions (awareness or discrimination); second, you must be able to translate your understanding into effective communication. If your accurate perceptions remain locked up inside you, they serve little purpose. However, if the language you use and your manner of communicating alienate the person with whom you are speaking (that is, if you communicate accurate perceptions poorly), effective understanding will not be communicated. For example, if you see that confrontation or challenge might help the person with whom you are speaking, but then confront him in such a way that he feels accused and punished, your lack of know-how will defeat an otherwise good intention. In both Phase I and Phase II, you will be trained in the know-how of core interpersonal skills.

3. *The assertiveness dimension.* High-level awareness and excellent know-how are meaningless unless you have the courage to use them when they are called for. Obviously, to be assertive without awareness and without know-how is only to court disaster. Human-relations-training experiences have seen too many participants launch into confrontations that are poorly timed, inappropriate, and poorly executed—attacks instead of invitations. Raw assertiveness is not a valuable commodity in interpersonal relating. However, it happens quite often (much too often) that participants with good perceptive abilities and a solid grasp of the technology of a variety of skills fail to use these resources adequately during the life of the group. For instance, some participants keep excellent logs, but then fail to translate these logs into useful interactions within the group. Assertiveness is the ability to take *reasonable* risks. Therefore, part of your training will focus on encouraging you to take these reasonable risks.

Being skillful means being able to orchestrate all three of these dimensions in actual interpersonal interactions. The word "skill," then, refers to a complex process. However, just as an athlete acquires a skill slowly and through steady and systematic practice (provided that he or she has some basic aptitude), you can acquire and strengthen interpersonal skills.

Chapter 3

The Skill of Self-Disclosure

Appropriate self-disclosure is a social skill that many people find difficult to learn. Some people are overdisclosers; that is, they either talk too much about themselves or talk too intimately about themselves in social situations that do not call for that kind of intimacy. Others are underdisclosers, unwilling to let others know anything about them except what can be picked up from observation. They speak little or not at all about themselves. They do not speak intimately of themselves even when the situation calls for intimacy. Self-disclosure can be seen as a continuum, with the overdisclosers at one end and the underdisclosers (or nondisclosers) at the other. Appropriate self-disclosure, then, is the "golden mean." I will present some background material here to try to help you place self-disclosure in the context of interpersonal relating and then try to present some guidelines for making your own self-disclosure, both in the group and in other appropriate interpersonal situations.

SELF-DISCLOSURE IN PERSPECTIVE

Since you are entering a laboratory experience that will require some self-disclosure of you, it may help to place the issue of self-disclosure in a wider context. This background material does not offer a definitive answer to the "problem" of self-disclosure; it is intended only to help you to think about this issue, to ask yourself relevant (or, as someone has said, even "impertinent") questions, and to begin to formulate your own values in regard to self-disclosure. Let's look at the issues more generally first and then apply them to our laboratory.

Openness Versus Privacy

Maddocks (1970) believes that we are being drowned in a flood of self-disclosure.

> People today tell complete strangers things they once wouldn't have confessed to a priest, a doctor or a close friend: their cruelest fears; their most shameful inadequacies; their maddest fantasies. We are witnessing something like the death of reticence [Maddocks, 1970, p. 50].

He sees everyone talking at once, but no authentic dialogue. Reticence, for him, is the art of knowing what can be said and what can't.

Bennett (1967), on the other hand, sees the dangers of an overemphasis on privacy.

> Privacy . . . is a graceful amenity, generally to be fostered, but with discriminating restraint and with due recognition of obligations as well as privilege. It is the writer's contention that the moral imperative is more often allied with the surrender of privacy than with its protection. . . . Secrecy within the community is incompatible with cooperation, inimical to the welfare and progress of the ingroup. . . . Strictly speaking, of course, sex is not ordinarily a private experience, but a peculiarly delicate and intimate transaction between at least two people. I submit that even in this sensitive area, more serious problems stem from mismanaged communication about sex—partners who cannot discuss it, children who must not be told, and alienation of the deviate—than from mere breaches of privacy. . . . The confessional is also respected as a confidential relationship. It should be noted that this, too, is a communication; a revelation, in fact, of the most private secrets to at least one other person. . . . The reference, in many religions, is to public confession. The Protestant sinner must bear witness "before men" to achieve absolution. Indeed, it is recognized in Catholic circles that the traditional confessional, intent on making peace with God, leaves unresolved the problem of making peace with the community. . . . The readiness of people to discuss their personal problems with neighbors, and even strangers, makes one wonder, in fact, whether confidentiality is so necessary to the privacy of the patient as the comfort of the therapist. There are therapists who believe that the therapeutic process is facilitated in the presence of an audience. The popularity of group therapy reflects a similar assumption that patients find help in sharing personal problems—that confession is good for the psyche as well as the soul. . . . The contemporary concern over privacy parallels a pervasive need to communicate. . . . Our dilemma will not be resolved by hiding away from each other in separate caves, but through more and more interpersonal communication, better managed. . . . The critical problem we face is not how to keep secrets from each other but how to facilitate this readiness to communicate. The overriding question is how to maintain an atmosphere of trust and confidence which will enable us to talk about personal affairs . . . freely. . . . It is the writer's conviction that the importance of honest communication in our interdependent relationships outweighs the sanctity of privacy as a social value. . . . Anyone who

> undertakes to influence the lives of other people must accept an obligation to let them know where he stands, to reveal his motives, to share his purposes [Bennett, 1967, pp. 371-376].

The issue, however, seems to be misstated when it is put in terms of openness *versus* privacy. Rather, there are times to be open and times to be private. Furthermore, there are both appropriate and inappropriate ways and degrees of being both open and private. Let's explore these issues a little further.

The Possible Dangers Of Concealment

Both Jourard (1968, 1971a, 1971b) and Mowrer (1964, 1968a, 1968b, 1973a, 1973b) offer extensive discussions, based on both theory and research, indicating that people who cannot reveal themselves appropriately run the risk of impoverished interpersonal lives and a wide variety of neurotic disorders. Although their findings are not definitive, they are stimulating and should not be ignored by anyone interested in his or her own interpersonal growth.

Although Mowrer (1973a) freely admits that psychological disorders can arise from genetic, biochemical, and ecological factors (in any combination), most of his work centers around people who get into psychological trouble because of what he calls "educational" or "decisional" factors. He believes that a great many people suffer from symptoms of psychopathology because of "bad choices and bad habits."

> Our major and overriding supposition is that the suffering individual . . . is uncomfortable ("neurotic") because he has been behaving dishonestly, irresponsibly, and without proper emotional involvement with and concern for other persons [Mowrer, 1973a, p. 527].

Treatment, then, is an educational process geared to helping the sufferer become honest (with both himself and others), responsible, and involved. The sufferer makes the problem worse by trying to conceal (from himself and others) the fact that he is acting irresponsibly. Deception has become a way of life, and this refusal to face up to "what is" is at the root of his disorder. Such concealment is a break with reality, and breaking with reality is the warp and woof of emotional disorder. The first step toward emotional reintegration is confessing (self-disclosing) one's dishonesty, irresponsibility, and self-centeredness to responsible and significant others.

Jourard (1968, 1971a, 1971b) has been investigating the consequences of concealment in a wider and perhaps more positive context (less moralistic and with a self-actualization orientation) than Mowrer's. Jourard believes that psychological disturbances often arise from stress or poorly managed stress. Overinvestment in privacy and concealment is, according to Jourard, very stressful.

> Every maladjusted person is a person who has not made himself known to another human being and in consequence does not know himself. Nor can he be himself. More than that, he struggles actively to avoid becoming known by another human being. He works at it ceaselessly, twenty-four hours daily, and it is work! [Jourard, 1971b, pp. 32–33.]

From a positive point of view, Jourard sees responsible self-disclosure as a symptom of personality health and a means of ultimately achieving a healthy personality. Being an appropriate self-discloser is part of the process of self-actualization.

> When I say that self-disclosure is a symptom of personality health, I mean a person who displays many of the other characteristics that betoken healthy personality will also display the ability to make himself fully known to at least one other significant human being. When I say that self-disclosure is a means by which one achieves personality health, I mean that it is not until I am my real self that my real self is in a position to grow. . . . People's selves stop growing when they repress them. . . . Alienation from one's real self not only arrests personal growth; it tends to make a farce out of one's relationships with people. The crucial "break" in schizophrenia is with *sincerity,* not reality [Jourard, 1971b, p. 32].

Perhaps neither Mowrer nor Jourard definitively proves that concealment in itself causes psychological disturbances, or that self-disclosure is such a primary element in emotional integration and self-actualization. But these hypotheses are supported by a wide variety of evidence that seems to demand, on the part of anyone looking for greater self-actualization in his interpersonal living, the kind of experiential examination of self-disclosure that is part of laboratory learning.

Cultural Blocks to Self-Disclosure

Significant self-disclosure may not be easy even under the best of circumstances, but our cultural and societal backgrounds against which it takes place make it even harder. Consider the following possible obstacles.

Dishonesty. Even though most Americans want honesty to be a primary value for their children (Kohn, 1959), lying, coverup, and dishonesty are daily facts of American life. Politicians in the highest offices of the land feel free to lie to everyone in order to advance their own causes. One begins to wonder whether American honesty is more myth than fact. It follows that you may begin to wonder: Whom can I believe in this group? Will others believe me?

Advertising. Many consumer products are not highly competitive with one another in price (for instance, soaps, automobiles), and therefore they allot substantial advertising budgets to emphasize differences that are less than substantial. We thus end up paying for large advertising budgets in order to deceive ourselves.

> The relaxed attitude veracity (or mendacity, depending on the point of view) and its complement, pecuniary philosophy, are important to the American economy, for they make possible an enormous amount of selling that could not take place otherwise. . . . One of the discoveries of the 20th century is the enormous variety of ways compelling language to lie. . . . We pay intellectual talent a high price to amplify ambiguities, distort thought, and bury reality [Henry, 1963, pp. 49; 91].

Henry suggests that most of us are not obsessive truth-seekers. We are often willing to let absurd or merely ambiguous statements pass (also see Schwartz, 1974).

Fromm (1947) has suggested that people in a success-oriented consumeristic society see their persons and their personalities as commodities to be marketed; and one can only wonder how much deceit, falsification, and façade are part of such a marketing process. Therefore you may ask: Will people in my group share themselves with me, or will they try to sell themselves? Am I out to share or to market myself?

A pervasive psychologism. For good or ill, psychology has made its way into the marrow of our culture. Overzealous in our belief in the scientific method, we set about "proving" what at times we might merely point to. To find out what a person is really like (so that we can hire, fire, commit, change, or otherwise control him), we must subject him to a battery (an apt word) of tests. The underlying assumption is that psychological truth is so hard to get at that the search must not be left up to amateurs. Not only must psychological truth be dug out with "scientific" instruments but, once it is excavated, it is often too dangerous (or abstruse) to communicate to the human quarry. There is a tendency, therefore, for such information to be used to control the person being tested (for his own good, of course; see London, 1969). Psychoanalysis, which in one "pop" form or another pervades our culture, suggests that a person cannot effectively mine the truth within himself without the assistance of a medically trained and highly paid expert in a long, tedious process. In the face of such psychologism, you may well ask yourself: Can I know myself well enough to make any self-sharing I may do here meaningful? How valid will what others say about themselves be? I have no quarrel with the legitimate uses of psychology, including psychological testing. I do have problems with psychology when it oversteps its bounds and degenerates into psychologism. Psychology in the best sense is an expansion of common sense, not a replacement for it. It makes sense to rely first on your own human experience and to make use of psychological "instruments" only as they are needed to expand and deepen and clarify this experience.

Sexism and stereotypes. The person who exhibits strength by suffering in silence has become a cultural stereotype in our society. "Little boys don't cry" is an early version of the masculine ideal; and the woman who, in the fiction of radio, TV, or the novel, confesses "I simply have to talk to someone" is thought to be confessing not a deep human need but her own weakness (even though such weakness might be understandable and even excusable in a woman!). If self-disclosure is not seen as weakness, it is seen as exhibitionism and, as such, as a sign of illness rather than of a desire for human communication. Very often the adolescent, in his discovery of himself and the "other," engages in a good deal of self-disclosure. But this drive to exchange intimacies, even though at times it may have overtones of exhibitionism or other kinds of problem behavior, should be regarded as a potential source of strength—being at ease in discussing oneself with significant others at an intimate level. Instead, it is often considered naïve and immature behavior. It is thought that such behavior will pass, just as the "natural neurosis" of adolescence passes.

When the adult finds it necessary to communicate thoughts to a friend at an intimate level, he often feels that he needs an excuse to do so. The person with a painful and perplexing personal problem is often loath to ask a friend to share it, and his friend is loath to encourage him to talk about it. The communication is difficult for both of them, for there is little cultural support for what they are doing.

Family background. Some families seem to encourage self-disclosure, whereas others inhibit it. People who are capable of high levels of self-disclosure generally believe that their parents were more nurturant than the parents of those who have difficulty with self-disclosure (Doster & Strickland, 1969). The point is that you learn self-disclosure and other behavioral patterns from your family. Intimate self-disclosure does not seem to be the norm in most American families. What were the norms for self-disclosure in your home? To whom do you disclose yourself most intimately?

Sources of Resistance within the Group Member

The impact of any of these societal or cultural factors, or any combination of them, may or may not influence you, even in a preconscious way, to be hesitant in disclosing yourself. Another factor inhibiting self-disclosure, as Maddocks (1970) points out, is a feeling that there are a growing number of overdisclosers in our society, who tend to inflict their own "tyranny of openness" (Altman & Taylor, 1973) on others. Certainly, the current human-potential movement (see Howard, 1970) in general espouses greater openness with the significant others in our lives, and there are those who fear that the individual's need for privacy may be in jeopardy. Simmel (1964) takes a different tack. He says that overdisclosers "lapse into matter-of-factness which no longer has any room for surprise" (p. 329). Discretion in self-disclosure prevents boredom.

Not all factors tending to inhibit (or encourage) self-disclosure are societal or cultural. There are sources closer to home, *intra*individual factors. Let's explore some of these.

The flight from self-knowledge. Self-disclosure is one of the principal ways of communicating not only with others but also with oneself. Perhaps, logically, the latter even comes first. An intriguing hypothesis is that many of us flee self-revelation because we are afraid of closer contact with ourselves. It has been said that the human organism seems capable of enduring anything in the universe except a clear, complete, fully conscious view of himself as he actually is. Self-disclosure crystallizes aspects of the self that a person would rather live with silently—however painful the living—than face. At least in this respect, then, *a group is only as threatening to a participant as he is to himself.* Inevitably, the individual is his own severest judge.

Much is being written about alienation and about identity conflicts, with attempts to establish both the social conditions and the intrapersonal dynamics of these problems. A person's flight from himself is, in large part, a flight from communication with himself. Self-alienation is frightening, but any kind of intimate contact with the problem self is seen as even more frightening. Thus self-alienation becomes self-reinforcing, its reward lying in its being supposedly less painful than its alternative. Even when a person gets out of contact with himself and with others to the degree that he flees to a mental hospital, this is still no guarantee that he is ready to face himself. Time and again in mental hospitals, when a patient is faced with the choice between the pain of alienation and the pain of therapy, he chooses the former, unable to find the courage to be. It is because a similar dynamic appears to operate in creating the psychopathology of the average that self-disclosure is stressed in laboratories in interpersonal relations.

The fear of intimacy. It is difficult to reveal oneself on a deep level to another without creating, by the very act of self-revelation, some degree of intimacy. In a group situation, for some reason, this intimacy seems to have a special intensity. The participants in various growth-group experiences are aware of this; and, even though they may have the courage to let others see the "mystery of iniquity" or even the "mystery of goodness" that they are, they fear the intimacy that such an act begets. They do not flee self-revelation as such. They flee intimacy. For some people, we are told, the fear of human relations is greater than the fear of death. They prefer, therefore, to skirt self-revelation and avoid intimacy. Some engage in sporadic acts of pseudo-self-revelation leading to pseudo-intimacy, in a games approach to human relationships. Others merely refuse to reveal themselves in any significant way. Their message, usually more nonverbal than verbal, is "Don't probe."

In sum, failed intimacy is a significant dimension of the psychopathology of the average. One way of counteracting this flight from intimacy is to engage in self-disclosure that is proportioned to the situation and that involves reasonable risk-taking. This is what is expected in the laboratory group.

Flight from responsibility and change. In some cases, flight from self-disclosure is a flight from responsibility—a flight from the anxiety and work involved in constructive personal change. Self-disclosure leads to the revelation of areas of deficit and of aspiration in human living. It is relatively easy to avoid both of these areas in day-to-day life. However, once a person declares what he finds unacceptable in himself and what goals he thinks he should be pursuing, he commits himself to change, and his avoidance behavior consequently becomes more painful. Self-disclosure commits one to conversion, to the process of restructuring one's life; it demands that one leave the security of his own house and journey into a foreign land, and most people balk at that. If one senses that conversion is impossible, he must avoid self-disclosure. Thus it is assumed that some people fear, or even downgrade, self-disclosure because of the behavioral consequences it entails. If the self-revelation takes place in a group, then the pressure to change is even greater than in a one-to-one situation, for the discloser must face the pressures and demands of a community.

The reverse halo effect. Another source of fear of self-revelation can be termed the "reverse halo effect." The "halo effect" refers to the fact that a person judged to be competent in one area is likely to be judged by many to be similarly competent in other areas (even when he is not). For instance, if a person is an expert in psychology, his aura of expertise tends to spread to other areas, and either he or others begin to look upon his pronouncements in theology or political science with the same awe. The reverse process sometimes stands in the way of self-disclosure in growth groups. The group member fears self-disclosure because he usually thinks first in terms of disclosing the worst in himself. If he tells the other members about incompetence in one area of living, he feels that they will assume similar incompetence or irresponsibility in related or even unrelated areas. If a person admits problems in his private life, he fears that others will assume incompetence in his professional life. This is especially true if the person's profession is closely related to interpersonal living, as is, for example, psychology. There are several ways of handling such a problem in the group. The participant can try to give a balanced view of himself, speaking alternately of strengths and weaknesses. Or the group can take up the problem of the reverse halo effect and discuss it directly. The reverse halo effect is related to the problem of stereotyping and categorizing. No one likes to be dealt with as a "problem," but there is a tendency in groups to identify the participants with their problems in living, for it is easier to deal with problems than with people.

The problem of guilt and shame. In a study dealing with the benefits of counseling (Talland & Clark, 1954), clients judged the therapeutic value of 15 topics discussed during counseling. There was general agreement on the relative value of the topics. Ratings showed a high correlation between the perceived helpfulness of a topic and its disturbing qualities. The topic called "shame and guilt" was experienced as extremely upsetting, but the discussion of this area during the sessions was considered to be very helpful.

"Shame" is a rich word and the experience of shame is, in many ways, a much richer experience than most of us have been led to believe. Lynd (1958) has written a remarkable analysis of shame and its relationship to identity. The root meaning of the word is "to uncover, to expose, to wound," and therefore it is obviously related to the process of self-disclosure. Shame is not just being painfully exposed to another; it is primarily an exposure of self to oneself. Particularly sensitive and vulnerable aspects of the self, especially in one's own eyes, are exposed. Shame has the quality of suddenness. In a flash one sees his unrecognized inadequacies, without being ready for this revelation of self to self; much less is one ready for exposure to the eyes of others.

Kaufman (1974) relates shame more directly to interpersonal living. He suggests, for instance, that shame may become an inevitable experience if a person's needs are not responded to appropriately by a significant other. The establishing of deeper relationships, then, leads to the possibility of generating shame.

> Relationships form when one person actively reaches out to another and establishes emotional ties. The process is one of creating a bond. The emotional bond that ties two individuals together is the interpersonal bridge between them. Such a bridge involves trust and makes possible experiences of vulnerability and openness between individuals. The bridge becomes a vehicle to facilitate mutual understanding, growth and change. These processes are disrupted whenever that bridge is broken.
>
> The interpersonal bond between individuals can be severed emotionally and the bridge broken without ever seeking actually to end the relationship. This emotional severing between people in an ongoing relationship is dynamically most relevant to inducing shame. The impact of shame increases profoundly when the relationship between the individuals concerned is of central importance to them. . . .
>
> If two individuals become bonded together, how and why would one of them sever the bond in such a way as to leave the other with a sense of being somehow not quite good enough? [Kaufman, 1974, p. 570.]

Kaufman suggests that, even when the "interpersonal bridge" is broken by one person and not restored, it can be restored by another person. This process of restoration involves recognition, rather than denial, of the fact that the person who feels shame experiences himself as a bad person in some way.

> Attempts at . . . ignoring those core beliefs (that the other is a flawed person, "bad" in some ways), convincing him otherwise, or trying to rid the person of

> them backfire. Such attempts deny the reality of those feelings and thereby engender shame about having them in the first place [Kaufman, 1974, p. 573].

Feelings of shame have to be recognized, not avoided or tampered with. This recognition begins the process of restoring the interpersonal bridge. As Lynd (1958) puts it:

> It is possible that the experiences of shame if confronted fully in the face may throw an unexpected light on who one is and point toward who one may become. . . .
>
> If . . . one can sufficiently risk uncovering oneself and sufficiently trust another person to seek means of communicating shame, the risking of exposure can be in itself an experience of release, expansion, self-revelation, a coming forward of belief in oneself [Lynd, 1958, pp. 20; 249].

Sometimes, Kaufman says, the person breaking the interpersonal bridge can reestablish the bond by revealing his own imperfect humanness and owning his part in making another feel shame. "The growth impact of having someone take that risk . . . is far greater than if he or she had never triggered off a shame experience in the first place" (Kaufman, 1974, p. 574). However, the significant other involved in the original shame-inducing experiences need not be the one who restores the bridge. This can be a new significant person—for instance, a friend or a therapist.

Fear of rejection. Underlying many of the reasons for not revealing ourselves very deeply to others is a fear of rejection. Many people think, at least subconsciously, "If others *really* knew me, they wouldn't accept me." This fear of rejection is often related to the person's feelings about himself. For instance, I fear that you won't accept me because, deep down, I don't accept myself in many ways. This common phenomenon underlines the fact that a climate of trust must be developed in the group if the members are to confide in one another. The ways of establishing trust are treated in Part III, The Skills of Responding.

Perhaps by now you are wondering whether you or anyone else can overcome the problems standing in the way of the kind of self-disclosure that is essential to a human-relations-training laboratory. The answer lies to a great extent in one word—"appropriate." If your self-disclosure in the group is appropriate, most, if not all, of the problems raised in the last few pages can be handled. What, then, are the criteria for appropriate self-disclosure?

APPROPRIATE SELF-DISCLOSURE

Self-disclosure is not an end in itself. Therefore, it is necessary to establish criteria for determining the appropriateness and relevance of self-disclosure in interpersonal situations. These criteria should be applicable both to your interactions within your training group and to your interpersonal life outside the group. Some of the criteria for determining the appropriateness of self-disclosure are

- *breadth:* the amount of information disclosed
- *depth:* the intimacy of information disclosed
- *duration:* the amount of time spent in disclosing oneself
- *target person:* the person or persons to whom information is disclosed
- *the nature of the relationship:* intimate friends, close friends, acquaintances, work associates, and so forth
- *situation:* the conditions under which the disclosure is made

Obviously, various combinations of the dimensions listed above are possible. For instance, a person who is constantly talking about himself (duration) and who reveals a great deal (breadth) of superficial information (depth) to almost anyone (random target person) on almost any occasion (situation) is usually called a "bore" and shunned.

Cozby (1973) approaches the question of appropriateness in self-disclosure by hypothesizing the differences between the self-disclosure of well-adjusted people and that of poorly adjusted people. The well-adjusted person, he suggests, engages in relatively high-level or significant self-disclosure, in terms of both breadth (amount) and intimacy (depth), to a few significant others (spouse, very close friends), while being moderately self-disclosing to others in his immediate social environment (ordinary friends, social acquaintances). "Moderate" self-disclosure means that the well-adjusted person will disclose enough to establish meaningful social bonds but not enough to be threatening or offensive. On the other hand, the poorly adjusted person either overdiscloses or underdiscloses (in terms of both breadth and depth) to virtually everyone in the environment. Poorly adjusted people have difficulty adapting self-disclosure to different kinds of target persons and different situations. The husband who says little about himself or what he does (breadth) to his wife (target person), finds intimate disclosure (depth) to almost anyone impossible, and cannot share himself even moderately when his wife shares herself deeply with him (situation) would fall into this category.

Working Criteria

Given this background, let's elaborate some working criteria for self-disclosure, both in the laboratory and in life itself. Some of the following criteria are adapted from Luft (1969). As you will notice, these criteria are interrelated, and therefore there is some overlap among them.

1. Goal-directedness. Know why you are disclosing yourself; that is, don't let self-disclosure become an end in itself. Self-disclosure for its own sake inevitably becomes exhibitionistic (if it is intimate and dramatic) or boring (if it is superficial and matter-of-fact). The overriding goal of self-disclosure is the fostering of relationships. Since in the laboratory you are trying to establish and develop relationships, your self-disclosure should be attuned to this goal. It should help you make contact with the other members of the group. Consider the following example:

> "I am a shy person. I find it hard talking to strangers and talking in groups. I get tongue-tied and forget what I want to say. At times I am so anxious that I feel I can't even enter an ongoing conversation. Since you are a group of strangers, I find it difficult getting going. I think it's only fair to let you know how I feel."

Since the goal of the group is to establish relationships, this person reveals an obstacle to that goal so that he can get the support he needs to go about his task.

2. *An eye to proportion.* The intimacy (depth) and the amount (breadth) of self-disclosure should be proportioned to your goal. If your goal is to establish a very close, ongoing, mutual relationship, your disclosure should eventually be quite intimate and extensive. In the group, however, even though your goal is to establish and develop relationships of some closeness, many of your fellow group members will probably not be your intimates. However, it is up to each of you to determine, in the flow of the group interaction, how close you would like these relationships to be and therefore just how much you would like to disclose yourself (in terms of both breadth and depth). The goal of the group is *not* dramatic self-disclosure (secret-dropping). It is not a place where you drag out the interesting skeletons in your closet and encourage others to do the same. This kind of disclosure is simply not proportioned to the goals of the group. On the other hand, merely superficial disclosures about yourself and your interpersonal style are not proportioned to the goals of the group either. The assumption is that you want to live interpersonally with some intensity, even in this time-limited group experience. Such intensity is impossible unless you are willing to reveal yourself—to risk some degree of intimacy with your fellow group members. In sum, it is impossible to set down exact rules on just how much you should reveal about yourself and how intimate your revelations should be. However, the remaining criteria of appropriateness should help you make realistic decisions.

3. *Respect and caring.* It is difficult to disclose yourself to people whom you do not care about or respect. You must have some basic respect for and desire to relate to the other members of your group; then disclosure becomes a means of developing and improving your relationship with them. This means that you must give others a chance. All of us have various interpersonal prejudices on an emotional level (even when these prejudices do not exist on the level of thought or will). My emotions may tell me that I do not really like to relate to older (or younger) people, people who seem too straight

(or too far out), ministers, nuns, and priests (or lay people), blue-collar people (or professionals), people who are not very bright (or too bright), people who are too emotional (or too cold)—the list could go on and on. Don't let these unexamined prejudices get the best of you from the beginning. More positively, work at respecting each of your fellow group members from the beginning of the lab; let your self-sharing be a sign of your respect.

4. *Ongoing quality of the relationship.* Self-disclosure is appropriate when it is not just a random act (revealing intimacies about yourself to the person sitting next to you on a plane or train). You should have some continuity of association with the target person or persons. Again, there will be a difference in terms of breadth and depth of self-disclosure between the time-limited ongoing relationships in the laboratory group and the relatively more permanent relationships of everyday life. On the other hand, your self-disclosure in a time-limited group need not be superficial. If you and your fellow group members establish a climate of mutual caring, cooperation, support, and trust, your disclosures may well be deeper in the group than in the everyday ("real-life") interpersonal contacts for which these supportive conditions do not exist. Thus what happens within the group points the way to what is possible outside the group. What happens in the group is real, but it is preparatory for day-to-day living outside the group.

5. *Mutuality.* Although Cozby (1973) notes that much of the research in self-disclosure up to this point is rather inconclusive, nevertheless some consistent patterns have been found. One is that, if you reveal yourself to others, others tend to reciprocate. Indeed, if such mutuality does not develop, self-disclosure is not appropriate. This is not to imply that self-disclosure in groups is a kind of tit-for-tat game, a kind of mindless exchange of intimacies (as promoted by certain "sensitivity games" that have been marketed in the past few years). Neither is your self-disclosure a way of extorting similar kinds of disclosure from your fellow group members. This lab experience will be successful when characterized by mutual, freely given self-disclosure. Although it is true that you and your fellow group members help one another to grow, you are not in the group precisely to help. In a helping relationship, one person (the client) usually reveals himself rather intimately, whereas the other (the helper) does not (at least not to the degree or in the way in which the client reveals himself; see Egan, 1975b, pp. 151–156, for a further discussion of this point). The revelations usually deal with the problems in living the client is experiencing. In this lab, however, self-disclosure is mutual and not necessarily problem-centered (although problems of interpersonal living are not excluded). Therefore, your group experience may well be therapeutic, and you and others may well be helped in a variety of ways; but human-relations training is not therapy, and helping is not its primary mode of interaction.

6. *Timing: Self-disclosure as emergent.* Self-disclosure should not be, in the poet's term, a "purple patch" sewn on your dialogue. It should emerge from and relate to what is happening in the group. For instance, suppose that

the members of a group run into one of those aimless periods when nobody seems to know what to say. The transactions that are taking place are relatively superficial and generate no significant discussion. During a lull, a member says:

> We seem to be having a hard time making deeper contact with one another—or at least I am. I think I know why I'm having difficulty. I haven't yet decided how intensively I want to live in this group. I'm not sure how intensively I want to live with my friends outside this group. Relating intensively here or outside demands a lot of work, and I'm not sure I'm up to it; but I'd like to talk about it more concretely here with you.

This participant voices his misgivings at a time when the group itself is drifting. His disclosure hits the mark because it emerges from what is happening (or not happening) in the group.

Self-disclosure can be disruptive if it is not geared to *making* something happen in the group. For instance, if John and Jane are trying to make concrete to each other the reasons they find it difficult to talk to each other in the group, George does not help at all if he begins to disclose his fears of speaking before large groups.

7. *The here and now.* Self-disclosure, even when it does not deal with what is happening in the here and now, should in some way be related to the here and now. In the lab you are trying to examine (and improve) your interpersonal style through the process of establishing and developing relationships with your fellow group members. This process is centered in the group and its transactions. It is not centered in any primary way in what takes place outside the group. This does not mean that you may not talk about your life outside the group. It does mean that, when you do talk about matters from the past or matters outside the group, you should take pains to relate what you are talking about to what is happening within the group. Here are some examples of how there-and-then material can either distract from or relate effectively to the here and now of the group interaction:

Unrelated self-disclosure	*Related self-disclosure*
My dad and I don't get along. He's usually down on my brother, too. It makes living at home difficult. Sometimes I just feel like getting an apartment on my own.	I'm reacting to you just the way I do to my dad. He doesn't listen to me; neither do you. I feel like turning my back on you, just as I feel like getting my own apartment. You're older, but I don't think that has anything to do with it.
I'm homosexual. I don't think society is fair to the homosexual. But then society is uptight about a lot of things. People have to fit	I think I trust you people enough to tell you that I'm homosexual. The reason I'm telling you is not to have you counsel me. I feel that some of you may reject me, but I'm hoping

in, to do the things everybody else does.	you won't. Since my sexual identification is something big on my mind, and since I fear rejection, I felt I had to bring it up.
I don't communicate much with my wife and children, but they let me alone.	I don't communicate much with my wife and children at home, and I'm falling into the same pattern here—except that you challenge me, whereas they don't. I should begin to challenge myself.
I have many fears. I'm afraid of strong men. I'm afraid of those who don't like me. I'm afraid I won't succeed.	I have many fears. For instance, I'm afraid of strong men, Bill, and I see you as strong. I fear rejection, and, Jane, I feel you don't like me. I'm afraid of failure, and I think I'm making a mess of this group experience. Well, it's out in the open now.

If the members of the group get taken up with the there-and-then concerns listed on the left, the group will suffer a loss of immediacy. Therapy groups often deal extensively with such there-and-then concerns; but mutuality, not problem solving, is the focus of the laboratory group in human relations.

8. Gradual self-disclosure. You should not overwhelm one another with your self-disclosure. As Luft (1969) notes, "What is revealed [should] not drastically change or restructure the relationships. The implication is that a relationship is built gradually except in rare and special cases" (p. 133). Because of the nature of laboratory learning, the process of establishing and developing relationships is usually accelerated; however, the process is still relatively gradual (see also Altman & Taylor, 1973). Self-disclosure at any stage must be supported by the level of trust within the group, and this trust emerges gradually. It must also be proportioned to the listeners' ability to assimilate the disclosure without being disoriented by it. That self-disclosure should be gradual does not mean that it should be used as an excuse for remaining shut up within oneself.

9. Reasonable risk. Just as a climate of trust enables you to risk yourself, it is also true that taking reasonable risks creates a climate of trust. Self-disclosure is a way of entrusting yourself to other people. A group in which all disclosures are overly safe soon becomes stilted and boring. On the other hand, if the risks taken in disclosing yourself are too high, others may become frightened and may respond unpredictably, as frightened people do. When this happens, the climate of support disappears and you are left out on a limb, exposed.

10. Impact. Try to judge what impact your self-disclosure will have on the other members of the group. I was once in a group in which one of the members disclosed some very intimate information about his past life in the very first meeting. Although his motive was to be open, he merely angered most of the participants by going too far too quickly. His disclosure was subsequently referred to as "the bomb," and it had the effect of inhibiting self-disclosure in others. As a result, the group remained at a relatively superficial level of self-disclosure throughout its life. It may well be that a number of the members were using memories of "the bomb" as a way of avoiding self-disclosure, but in any case the premature self-disclosure created a climate that aided and abetted such avoidance.

11. Shared context. Self-disclosure is appropriate when the members involved in giving and receiving the disclosures are working at the same problem; or, as Luft (1969) puts it, "the assumptions underlying the social context suggest that there is enough in common to sustain the disclosure" (p. 133). Therefore, if one of the members of the group was there principally to "observe" and to "learn about the group process" without seriously involving himself in the group interaction, there would be no substantial shared context between him and the other members. The contract stated in Chapter 2 helps assure group members that they have gathered together for the same general purposes and that they share certain underlying assumptions. In laboratory groups characterized by "goallessness," whether such goallessness is planned or not, it is almost impossible to count on the kind of shared context that promotes self-disclosure.

12. Response as a sign of reception and validation. The most productive immediate response to a person's self-disclosure is not someone else's self-disclosure but some indication that the disclosure has been heard. Self-disclosure demands some kind of response. Part III discusses the reasons why the communication of basic empathic understanding is ordinarily the best response. This kind of response validates reception of what has been disclosed. The discloser can say "Oh, someone has listened to and understood what I just said about myself." If a person's self-disclosure is met by silence, or is not acknowledged in some creative way, he or she is left hanging and will probably think twice before engaging in self-disclosure again. This whole question of response will be explored at length in Part III.

13. A balanced picture. For some, the term "self-disclosure" immediately evokes a vision of personal weaknesses. However, self-disclosure that involves merely the "confession" of weaknesses is inappropriate. If you are to come to a deeper understanding of your interpersonal style, you have to get in touch with your interpersonal strengths and resources (both those you are currently using and those you haven't yet mobilized). The laboratory is a time for confirming and developing strengths, not just a time for exploring weaknesses.

Self-disclosure can be viewed in terms of the D-, M-, and B-needs and functions referred to in Chapter 1.

- What interpersonal deficiencies do you experience? Examples of these D-problems are a fear of meeting people, a need to control or manipulate others, emotional prejudices toward various categories of people, and so forth.
- What M-involvements take up too much of your time? Some M-problems are spending a great deal of time with friends in superficial banter, remaining neutral in one's transactions with others because of not knowing how to assert oneself responsibly, and so forth.

What B-strengths and B-goals do I have? For instance, I find myself a caring person and want to develop counseling skills. Or I find that I have certain skills that enable me to mediate between others, although I seem to use this resource infrequently. Ultimately, confirming and developing strengths (B-functions) should predominate in group interactions. It is amazing how seldom people review their strengths as a way of encouraging and stimulating themselves.

These, then, are some criteria to follow in order to make your self-disclosures appropriate. Can you think of others? If these criteria are observed, most of the problems mentioned in the sections dealing with cultural and intraindividual sources of resistance to self-disclosure can be handled. Most objections to self-disclosure are to *inappropriate* self-disclosure.

THE VARIED MODES OF SELF-DISCLOSURE

The truth of the matter is that you are disclosing yourself constantly, whether you intend to or not, because your self-disclosure is in no way limited to the actual verbal revelations you make about yourself. There are at least three sources of self-revelation.

Verbal self-disclosure. You can use words to let others know either what is already in the public forum ("I'm a rather shy person") or something about yourself that is hidden from others ("I daydream a lot about being a successful musician and being liked by a lot of people"). Verbal self-disclosure of material that is already in the public forum (such as information about your interpersonal style) is an indication that you yourself are in touch with that information.

Nonverbal self-disclosure. Many nonverbal cues accompany your verbal behavior. For instance, if you are invited to dinner, your words may say that you will come, but the hesitation in your voice and the tenseness in your body say that you don't want to come. The manner in which a verbal message is communicated itself produces "messages." These messages include the speed, tone, inflection, intensity, and emotional color of the voice, along with such cues as eye contact, body stance, facial expressions, and gestures. Try these exercises. Choose a partner and ask your partner out to dinner. Have your partner reply both verbally and nonverbally in a way that says decidedly "yes." Now make another request of your partner and have your partner say "yes" verbally but "no" nonverbally. Repeat the process; but this time, respond to the requests yourself. There is one danger involved in awareness of nonverbal behavior: if you become too preoccupied with nonverbal behavior, you can take it out of context, exaggerate its meaning, and read your own projections into it. Therefore, in this group experience, try to get a feeling for your own nonverbal style. Try to catch yourself in discrepancies between what you are saying verbally and what you are saying nonverbally. Also ask others for feedback on your nonverbal style. You may discover that you are sending "messages" that you don't intend to send, or that at times there are discrepancies between your verbal and nonverbal messages.

Self-disclosure through actions. What I do also tells others a great deal about myself. If my actions are in the public forum, this kind of self-disclosure takes place in the public forum. What are some of the actions that reveal things about me?

- how I dress
- the kinds of friends I choose
- the kind of work I do
- the quality of my work
- what I do with my leisure time
- what I read
- what I watch on TV (if I watch TV)
- the kinds of learning experiences I pursue
- how I communicate in one-to-one situations
- how I communicate in small groups

Obviously, you could add many more things to the list. Notice, however, that much of this kind of behavior will not be known to the members of your lab group. It is up to you to let them know what kind of person is revealed through these actions. This does not mean that you should spend a great deal of time talking about what you do outside the group. Rather, if what you do outside the group says something about you that you would like to communicate to the members of your group (especially if it relates to your interpersonal style), you can tap this source of self-disclosure. Obviously, the way you act inside the group is seen by everybody. You need not reveal that verbally. Your behavior inside the group becomes the basis on which you receive feedback from your fellow group members.

EXERCISES IN SELF-DISCLOSURE

Before doing these exercises, you should recall once more that self-disclosure is not an end in itself, that it cannot take place all at once, that it should be integrated into your interactions with your fellow group members, and that it should help you pursue both your own personal goals and the group's goals. Exercises should help improve the quality of your participation within the group, but they do not substitute for your own spontaneity in the here and now of group interaction.

The following exercises stimulate you to think about yourself in various ways. They do not tell you how to use the information that comes up (it is up to you to make your disclosures appropriate). What you share and what you keep to yourself are up to you. The process involved in sharing (in pairs, in the entire group, or through a round robin) will be indicated to you in each instance by the trainer.

Exercise 4: The "I Am" Exercise

In this exercise you are asked to take the stem "I am . . . " and finish it in 20 different ways. Complete the stem in ways that will help you reveal yourself to the group more effectively.

For example, a group participant once came up with a list something like this:

- I am intelligent.
- I am fun-loving.
- I am an elitist.
- I am sexually immature.
- I am irresponsible at times, placing comfort ahead of carrying out principles I hold.
- I am very fickle in my likes and dislikes.
- I am shy and uncomfortable in situations where I feel threatened by those more intelligent and articulate than I.
- I am often enough quite bored with everybody and everything.
- I am witty and humorous.
- I am personable and enjoyable to be with.
- I am understanding, at least when I want to be.
- I am very open to new ideas and interpretations.
- I am a conservative iconoclast.
- I am an opportunist, and manipulative.
- I am more careful than careless.
- I am a person of conflicting themes.
- I am seductive in various ways.

Exercise 5: Incomplete Sentences as Stimuli for Self-Disclosure

In groups, it is sometimes difficult to think of what you want to say about yourself in order to begin establishing relationships with others. This exercise, then, can serve as a stimulus to help you think about your interpersonal life. First of all, finish each incomplete sentence. Do this relatively quickly;

that is, don't spend a great deal of time thinking of what you will (or should!) put down. Let the sentence stem act as a stimulus, and put down whatever arises naturally.

1. People who love me . . .
2. One thing I really like about myself is . . .
3. I dislike people who . . .
4. When people ignore me, I . . .
5. The way I express my generosity to others is . . .
6. When someone praises me, I . . .
7. When I relate to people, I . . .
8. When I don't like someone who likes me, I . . .
9. Those who really know me . . .
10. When I let someone know something about my shadow side . . .
11. My mother . . .
12. My moods . . .
13. I am at my best with people when . . .
14. When I am in a group of strangers, I . . .
15. I feel lonely when . . .
16. I envy . . .
17. When someone is affectionate with me, I . . .
18. When I take a good look at my interpersonal life, I . . .
19. The way I handle jealousy is . . .
20. I think I have hurt others by . . .
21. Those who don't know me well . . .
22. My brother . . .
23. The person who knows me best . . .
24. An important interpersonal value for me is . . .
25. What I am really looking for in my relationships is . . .
26. I get hurt when . . .
27. I daydream about . . .
28. My family . . .
29. When someone confronts me, I . . .
30. I am at my best with people when . . .
31. What I feel most guilty about in my relationships with others is . . .
32. I like people who . . .
33. When someone gets angry with me, I . . .
34. My sister . . .
35. Few people know that I . . .
36. When I think about intimacy, I think of . . .
37. When I meet someone who is very assertive, I . . .
38. When someone gets to know the best in me . . .
39. When I'm not around, my friends . . .
40. Most people think that I . . .
41. One thing I really dislike about myself is . . .
42. When I am with a group of my friends, I . . .
43. I get angry with another when . . .
44. What I distrust is . . .
45. One thing that makes me nervous in interpersonal situations is . . .

46. When I really feel good about myself, I . . .
47. When others put me down, I . . .
48. In relating to others, I get a big lift when . . .
49. Regarding relating to others, this year I learned that . . .
50. When I like someone who doesn't feel the same way about me, I . . .
51. I feel awkward with others when . . .
52. When others get parental with me, I . . .
53. The thing that holds me back in my relationships with others is . . .
54. I feel let down when . . .
55. When I share my values with someone, I . . .
56. I would like the person I marry . . .
57. Others like it when I . . .
58. Interpersonal relationships are important, but . . .
59. When someone sees the ways in which I am vulnerable . . .
60. In interpersonal situations, what I run away from most is . . .

Once you have completed the sentence stems, go back and circle the ones that have the most importance to you, say a great deal about you, and make you think about your interpersonal life in different ways, and that you would like to reveal to the members of your group. Use these sentences as the content of the interactions in which you practice the skills of communication.

Exercise 6: Mapping Some Dimensions of Your Interpersonal Style

This exercise is meant to help stimulate your thinking about your interpersonal style. Appropriate self-disclosure means (1) disclosure that is relevant to the goals of the group (establishing relationships and through this process coming to a better understanding of your interpersonal style) and (2) disclosure of what you yourself want to reveal. This exercise should help you become more concrete in your reflections on yourself and in your disclosure.

Read each pair of statements (*a* and *b*) and determine which of these statements describes you (that dimension of your interpersonal style) better. Next, examine the rating sheet on page 62. Once you determine whether *a* or *b* fits you better, then determine how good the fit is. For instance, if in 1 you choose statement *b* and think that statement *b* fits you very well, place an *x* in the column "very much like *b*." Do all 26 pairs and then read the process directions that follow the list of pairs.

Statements of Interpersonal Style

1. a. I'm shy. My shyness takes the forms of fear of meeting new people and of revealing myself to the friends I have.
   b. I'm an outgoing person. I enjoy meeting new people and seek out opportunities to do so.
2. a. I'm not assertive enough. Others can run roughshod over me and I just take it.
   b. I stand up for my rights pretty well. I am kind to others, but I don't let them manipulate me.
3. a. I get angry very easily and let my anger spill out on others in irresponsible

ways. I think my anger is often linked to my not getting my own way.

b. Although I become angry at times, I rarely lose control. When I'm angry, I seek out others and try to settle what's bothering me.

4. a. I'm a lazy person. I find it especially difficult to expend the kind of energy necessary to listen to and get involved with others.

   b. I'm a very energetic person. I like listening to and getting involved with others.

5. a. I'm somewhat fearful of persons of the opposite sex. This is especially true if I get the feeling they want some kind of intimacy with me. I get nervous and tongue-tied.

   b. I get along very well with persons of the opposite sex. I regard them as individuals, not objects, and I can relate with them at different levels, ranging from "casual" to "good friends" to "intimate."

6. a. I'm a rather insensitive person. I find it hard to know what others are feeling. I'm the bull-in-a-china-shop type.

   b. I'm usually aware of what others are feeling. Often I find myself experiencing something of the same emotion as someone else, just in listening to him.

7. a. I'm overly controlled. I don't let my emotions show at all, if possible. Sometimes I don't want them to show even to myself.

   b. I express my emotions well. I don't dump them on others, but I don't try to hide them, either. Others usually know what I am feeling.

8. a. I like to control others, but I like to do so in subtle ways. I want to stay in charge of interpersonal situations at all times.

   b. There is a great deal of mutuality in my relationships. I don't let others control me; neither do I desire to stay in charge of interpersonal situations. There is a lot of give-and-take in my relationships.

9. a. I have a need to be liked by others. I seldom do anything that will offend others, because I have a need to be seen as a good guy.

   b. Whether or not I am liked by others is important to me, but it doesn't get in the way of my doing what I think is right for me. I don't like to offend others, but I don't worry about getting others' approval.

10. a. I never stop to examine my own values. I think I hold some conflicting values. I'm not even sure whether I'm interested in relating deeply with others.

    b. I frequently examine my own value system, and when I find myself holding conflicting values I try to straighten things out, to determine my priorities. For instance, relating well with others is very important to me.

11. a. I feel almost compelled to help others. I get nervous when I am not engaged in helping. People with problems are almost necessary for me.

    b. When I help others, it's for them, not me. I suppose I get some personal satisfaction out of being altruistic, but I don't thrive on other people's problems.

12. a. I'm very sensitive, easily hurt. I send out messages to others that say "Be careful of me."

    b. I'm an easy-going person, in the sense that I'm not oversensitive to being hurt. I roll with the punches pretty well; I can laugh at myself. Others know that they can be loose when they're around me.

13. a. I'm a counterdependent person. I always have to show others that I'm free and an individual in my own right. I find it difficult to get along with others, especially those in authority.

    b. I'm an interdependent person. Although being an individual in my own right is a value for me, it is also very important for me to allow others to influence me. My friends and I influence one another because we want to.

14. a. I'm an overly anxious person, especially in interpersonal situations. But I don't know why I'm like that.
    b. I'm a relaxed person. Although I do get anxious at times, being relaxed in interpersonal situations is more characteristic of me.
15. a. I am, at least relatively, a colorless, uninteresting person. I'm bored with myself at times, and I assume that others are bored with me.
    b. I'm a colorful, interesting, dynamic person. Others enjoy my presence. I add life and excitement at gatherings, and I like this part of myself very much.
16. a. I take too many irresponsible risks in interpersonal situations. I'm rash and impulsive. I lack adequate self-control.
    b. I risk myself to some degree in interpersonal situations. The risks I take, however, are not foolish or irresponsible. I exhibit adequate self-control.
17. a. I'm stubborn and pig-headed. I'm very opinionated, and I'm ready to argue with almost anyone on anything. This characteristic puts people off.
    b. I have an open mind. Although I have ideas of my own, I don't go around looking for arguments; neither do I stick rigidly by my opinions in the face of other, perhaps more reasonable, views. I enjoy sharing views with others.
18. a. I'm somewhat sneaky and devious in my relationships with others. I seduce people in various ways (not necessarily sexual) by my charm. I get them to do what I want.
    b. I'm forthright and direct in my relationships with others. If I want something from them, I say so, as plainly as possible. I'm not sneaky or seductive.
19. a. I'm selfish and lazy, especially when it comes to responding to others. I put my own needs above the needs of others.
    b. I'm capable of self-sacrifice and discipline without being a martyr. I can put the needs of others ahead of my comfort.
20. a. I feel socially inept at times. I don't do the right, the human, thing at the right time. For instance, I don't notice when others are suffering, and as a result I seem callous.
    b. I'm socially adept most of the time. I'm sensitive to what those around me are feeling, and I usually respond to their feelings in a human way.
21. a. There is a degree of loneliness in my life. I don't think that others like me. I spend a great deal of time feeling sorry for myself.
    b. I'm a pretty secure person. Occasionally I experience loneliness, but when I do I'm able to keep things in their proper perspective. I seldom start feeling sorry for myself or think that everyone dislikes me.
22. a. I'm stingy, with money and with time. I don't want to share what I have.
    b. I'm a generous person. I like to share what I have with others—time, possessions, happiness, sorrow—and I like to receive what others have to share with me.
23. a. I feel a bit out of it, for I believe I'm inexperienced and somewhat naïve. When others talk about their experiences, I feel apprehensive or left out or find it hard to get a feeling for what they mean. I've lived too sheltered a life.
    b. I'm a talented person. I've been around, I'm socially aware, and I'm competent. I rarely get embarrassed when others talk about their experiences, for I know that I too have experiences worth disclosing.
24. a. I'm something of a coward. I find it hard to stand up for my convictions when I am opposed even slightly. It's easy to get me to retreat.

b. I'm a courageous individual who is unafraid when it comes to asserting himself. My beliefs and values require me to confront myself, so I'm not "at sea" when others question my actions or views.

25. a. I find it hard to face conflict between myself and someone else or to observe conflict between two others. I get scared. I run from it. I'm more or less a peace-at-any-price person.

    b. I have a lot of determination when it comes to working things out. I don't back out when it seems that conflict is inevitable. I don't like to avoid heavy issues, and I don't like to pretend that they don't exist.

26. a. When I am confronted, even legitimately and responsibly, I tend to attack my confronter and to respond in other defensive ways.

    b. When I am confronted, my tendency is to listen to what the other has to say, think about the validity of what he has said (from both my frame of reference and his, and to respond nondefensively.

This list is not exhaustive. It is meant to help stimulate your thinking about yourself in ways related to the goals of training.

After you have filled out the rating sheet, turn it on its side so that the *b*s are on top. As you have probably noticed, the *a* statements are negative, the *b* statements positive. If you draw lines connecting each *x*, your rating sheet becomes a kind of graph, the "hills" indicating your resources and the "valleys" your areas of deficit. From this graph, it is hoped that you can obtain a *balanced* view of yourself, seeing your areas of deficit in the context of your strengths and resources. As you move along in the group experience, you will have the opportunity—both as you watch yourself and as you receive feedback from your fellow group members—to determine *experientially* whether your own judgment of your strengths and weaknesses is accurate. An accurate view of resources and areas of deficit is essential if you want to embark on any meaningful change-of-behavior or change-of-style program.

- Where are the mountains? Where are the valleys?
- Do any major themes emerge?
- Are there a couple of *x* marks that stand out for you?
- Do you seem to be too hard on yourself? Too easy?
- Do you think you could use any of your identified strengths to help yourself get at any of the areas of deficit?
- Do you have a clearer idea of some of the things you may want to disclose to your fellow group members?
- What significant dimensions of *your* interpersonal style were not alluded to here?
- What elements would you like to change in your style? What elements would you like to confirm?

Make some of your insights the "content" of the skills-practicing sessions, and share those you think are important (and want to share) in the full group meetings.

Rating Sheet

	Very much like *a*	Fairly much like *a*	Somewhat like *a*	Somewhat like *b*	Fairly much like *b*	Very much like *b*	
1a.							1b.
2a.							2b.
3a.							3b.
4a.							4b.
5a.							5b.
6a.							6b.
7a.							7b.
8a.							8b.
9a.							9b.
10a.							10b.
11a.							11b.
12a.							12b.
13a.							13b.
14a.							14b.
15a.							15b.
16a.							16b.
17a.							17b.
18a.							18b.
19a.							19b.
20a.							20b.
21a.							21b.
22a.							22b.
23a.							23b.
24a.							24b.
25a.							25b.
26a.							26b.

Exercise 7: Self-Disclosure Cards

Review the last three exercises, all of which center on possible topics for self-disclosure within this lab experience. In reviewing these three exercises, write down the topics that pertain especially to you and your interpersonal style, one topic per card. Then, when it is time for you to talk about yourself in the skills-practice sessions, go through the cards briefly and choose a significant topic to talk about. This preparation will enable you to talk about yourself in substantial ways instead of choosing superficial topics on the spot. It will also enable you to control the intensity of your self-disclosure. For instance, as you become more comfortable in the group (as a climate of trust develops), you may choose to talk about deeper, more personal topics. As you move through the lab experience, you can both add new cards and develop the themes on the cards you already have. This exercise should help you make your self-disclosure both substantial and appropriate.

CHAPTER 3: FURTHER READINGS

On self-disclosure:

Bennett, C. C. What price privacy? *American Psychologist*, 1967, *22*, 371–376.

Cozby, P. C. Self-disclosure: A literature review. *Psychological Bulletin*, 1973, *79*, 73–91.

Culbert, S. A. *The interpersonal process of self-disclosure: It takes two to see one.* Fairfax, Va.: Learning Resources Corporation, National Training Laboratories, 1967.

Egan, G. Self-disclosure. Chapter 7 in *Encounter: Group processes for interpersonal growth.* Monterey, Calif.: Brooks/Cole, 1970. Pp. 190–245.

Jourard, S. M. *Disclosing man to himself.* New York: Van Nostrand Reinhold, 1968.

Jourard, S. M. *The transparent self* (rev. ed.). New York: Van Nostrand Reinhold, 1971.

Keen, S., & Fox, A. V. *Telling your story: A guide to who you are and who you can be.* New York: Doubleday, 1973.

Luft, J. *Of human interaction.* Palo Alto, Calif.: National Press Books, 1969.

Mowrer, O. H. Integrity groups today. In R-R. M. Jurjevich (Ed.), *Direct psychotherapy: Twenty-eight American originals* (Vol. 2). Coral Gables, Fla.: University of Miami Press, 1973. Pp. 515-561.

Powell, J. *Why am I afraid to tell you who I am?* Niles, Ill.: Argus Communications, 1969.

On guilt and shame:

Egan, G. *Encounter: Group processes for interpersonal growth.* Monterey, Calif.: Brooks/Cole, 1970. Pp. 213–231.

Kaufman, G. The meaning of shame: Toward a self-affirming identity. *Journal of Counseling Psychology,* 1974, *21*, 568–574.

Lynd, H. M. *On shame and the search for identity.* New York: Science Editions (Harcourt Brace Jovanovich), 1958.

Piers, G., & Singer, M. B. *Shame and guilt.* Springfield, Ill.: Charles C Thomas, 1953.

Chapter 4

Concreteness in Communication

Speaking vaguely or in abstractions creates distance between you and the person to whom you are communicating. Whereas it is not always possible to be completely clear in your communication with others (interpersonal transactions can, at times, be very complicated), still, obscurity of communication is hardly a goal (unless you are trying to avoid or manipulate someone). Concreteness helps greatly to make communication clear. In this chapter I shall discuss concreteness in terms of specificity of self-disclosure, including the disclosure of feelings and emotions. If self-disclosure is to be immediate and engaging, it must be concrete. Concreteness, then, is a communication skill, to be learned as other skills are learned. Concreteness means speaking about specific experiences, behaviors, and feelings.

Experiences. "Experience" here refers to what happens to me, what others do to me.

- I have a headache.
- John confronted me for taking up too much time in our last group meeting.

Note that experiences can also include thoughts and imaginings, especially when these arise without our willing them.

- The snide remark she made about me kept coming up in my mind over and over again.
- Whenever I just let myself daydream, I see myself as some kind of hero.

Behavior. "Behaviors" here means what I do or don't do, what actions I take or refrain from taking.

- I didn't tell Karen that she hurt me.
- I run away from situations in which I find myself getting hurt.

Thoughts and imaginings can also be behaviors if they are directly willed rather than merely allowed to happen.

- I keep thinking about how I will get my revenge for what John did to me in the group last week.
- I often picture myself being bold and assertive in front of my boss.

Feelings. "Feelings" refer to the emotions that accompany my experiences and my behavior.

- I'm depressed because I've had this headache for two days.
- I felt glad when I finally told Norm that he means a lot to me.

In exploring your interpersonal style, you are asked to be concrete in your exploration; that is, you are asked to talk about the specific experiences, behaviors, and feelings that relate to that style.

- I'm shy; that is, I get very scared when people tell me either verbally or nonverbally that they like me or want to get close to me. I find myself avoiding people after they let me know they want to establish a closer relationship with me. I'm afraid that this is going to happen in this group.

This person tells us something about his interpersonal style in a concrete way; that is, he talks about specific experiences, behaviors, and feelings. (Pick them out from his statement.)

Feelings, experiences, and behaviors can be *overt*—that is, others can see what you are feeling, experiencing or doing; or they may be *covert*—that is, hidden from view. Let's take a look at a few examples.

- *Overt experience:* John tells me he's angry when he gets angry with me, and that keeps the lines of communication open between us.
- *Covert experience:* My imagination goes wild with possibilities when I think of traveling with you this summer.

Overt experiences can be observed; covert experiences are invisible to the observer.

- *Overt behavior:* When you confront me as a way of dumping your anger, I clam up and don't respond to you at all.
- *Covert behavior:* I think about what happens in this group a great deal between sessions. I daydream about the kinds of interactions I'd like to have.

Again, covert behavior has to be revealed; it is not directly observable.

- *Overt feelings:* (heatedly) I resent your being so parental to me. You chide me, you take care of me. I want to be your equal here!

In this example, the speaker actually labels his feelings. A person, may, however, *express* feelings (and therefore the feelings are *overt*) without labeling them. Overt feelings, then, are feelings that are either talked about or expressed in some way.

- *Covert feelings:* I may appear calm on the outside, but whenever I'm confronted in this group, my gut begins to scream. It happened just now.

The assumption is that covert feelings are sufficiently screened or covered up so that the average onlooker does not pick them up.

Why do people fail to be concrete? One reason is that it is riskier to be concrete when talking directly to another person. It is too easy to hide under generalities and abstractions. For instance, which of the two statements below is more concrete? And which is riskier?

- I don't always have good feelings about myself, and this cuts down sometimes on my interpersonal involvement.
- I get depressed because I actually feel that I'm ugly—whether it's true or not. I think of myself as very bland. John, you're outgoing, personable, attractive. I let myself compare myself with you—and I get more depressed. So I'm quiet in here a great deal. Self-pity ties me up in knots and I can hardly speak.

Interactions that lack concreteness tend to be boring. Therefore, if you find yourself bored during the course of the group, ask yourself whether you are being concrete in talking about yourself, and whether others are being concrete. Another way to add concreteness to your interactions is to "own" what you say personally. There is a strong tendency, when you talk about yourself, to use substitutes for the personal pronoun "I"—such as "you," "one," and "people."

- *Not owned* (not concrete): It's not easy to talk directly about yourself.
- *Owned* (concrete): I'm afraid to talk about my shyness here. I'm afraid that people will come down hard on me and it will hurt.
- *Not owned:* People freeze up when confronted.
- *Owned:* When you told me I was swallowing my emotions, I practically collapsed inside, but I tried to put on a brave face.

When you use "I" instead of its more abstract substitutes, you force yourself to be more concrete and to own what you are saying.

EXERCISES IN CONCRETENESS

Exercise 8: Speaking Concretely about Experiences

In the following exercise you are asked to speak about some of your specific experiences—first vaguely, then concretely. Choose experiences that say something about your interpersonal style or are related to the training group experience. Study the following examples.

Example 1

- *Vague statement of experience:* People don't always understand me very well.
- *Turning a vague statement into a concrete one:* People tell me that I come across as cold and aloof. As a result, they don't make friendly overtures to me. They say it's hard to talk to me.

In this first example, the trainee talks about experiences that affect his interpersonal style.

Example 2

- *Vague statement of experience:* People sometimes don't let people finish things here.
- *Turning a vague statement into a concrete one:* Alex, just when I thought I was telling you something important, you interrupted and started talking about *yourself.* It was as if what I was saying didn't even matter to you.

This trainee is talking about a concrete experience in the training group.

Now write out four examples like the examples above—first a vague statement about your experience and then a statement turning the vagueness into something concrete. Remember to choose experiences that relate to your interpersonal style and that refer to what is happening to you in the training group. In this exercise stick to experiences—that is, *what happens to you.* In writing this and the following exercises on concreteness, imagine yourself speaking to your group or to particular members of the group. Don't just "make it up"; make it as real and as pertinent to your group as possible.

Exercise 9: Speaking Concretely about Your Behavior

In the following exercise you are asked to speak about some of your specific behaviors that relate to your interpersonal style or to what you do in the training group. Study the following examples.

Example 1

- *Vague statement of behavior:* I'm not so good at some of the challenging aspects of interpersonal behavior.

- *Turning a vague statement into a concrete one:* Although I think I'm fairly good at challenging others to look at discrepancies in their interpersonal lives, I act defensively when I am challenged. I too often say "Oh, no, that's not what I'm doing—let me explain." I seldom acknowledge failure or ignorance.

In this example, the speaker talks about behavior that constitutes part of his present interpersonal style.

Example 2

- *Vague statement of behavior:* People, including me, don't always live up to the contract in this group.
- *Turning a vague statement into a concrete one:* If I have difficulty relating to someone in this group, I often talk to him outside the group. I've had two or three of these relatively safe private talks.

This trainee speaks about concrete behavior that relates directly to the training group.

Now write out four examples like the examples above—first a vague statement about your behavior and then a statement turning the vagueness into something concrete. Remember to choose behaviors that relate to your interpersonal style and that take place in the training group. In this exercise, stick generally to behaviors—that is, *what you do or don't do.*

Exercise 10: Speaking Concretely about Your Feelings

In the following exercise you are asked to speak about some of your specific feelings that relate to your interpersonal style or to what happens to you in the training group. Study the following examples.

Example 1

- *Vague statement of feelings:* I'm kind of not sure what's going on here.
- *Turning a vague statement into a concrete one:* I'm really at sea right now. I feel like talking about myself, but I'm afraid of taking the group's time. I'm really on edge. Not knowing how you feel about me makes me very, very cautious.

Example 2

- *Vague statement of feelings:* It seems that things are funny between you and me, Joan.
- *Turning a vague statement into a concrete one:* It's hard for me to say this to you, Joan. I feel silly—more than silly—I feel *ashamed* when we talk. I'm like a little kid around you, one who feels he's done something wrong. I'm afraid of telling you about myself and how I feel.

Now write out four examples like the examples above. They should deal with feelings about yourself and how you relate to others, to specific people in your training group, or to specific experiences in the training group. Feelings accompany experiences and behaviors, but in this exercise *make the expression of feelings primary.*

Exercise 11: Speaking Concretely about Experiences, Behaviors, and Feelings Together

In the following exercise you are asked to "put it all together"—to speak concretely about yourself, your interpersonal style, and your experiences in the training group through statements describing experiences, behaviors, and feelings. Study the following examples.

Example 1

- *Vague statement:* Everything seems to be going wrong.
- *Turning a vague statement into a concrete one:* When the group first started, I was eager and optimistic. I was very active; I threw myself into the task of getting to know you. I contacted you a lot. I've had every intention of giving myself completely to establishing relationships. Then last week I was jumped on for being honest about my feelings toward Jack. I told you, Jack, that I thought you are running away from your feelings, and everybody rushed to defend you. *I* felt attacked. I haven't said much this week. I'm sitting here on edge, feeling something like an outcast. I want to get back in.

Example 2

- *Vague statement:* I'd like to hit it off better with you.
- *Turning a vague statement into a concrete one:* I like you. That's my first reaction. You seem intelligent and sensitive, and that attracts me. Right now I'm doing something I don't ordinarily do. Usually when I like someone, I play it cool, hold back, find out whether it's going to be mutual or not. With you I'm making myself vulnerable. You contact me in the group, give me feedback. You seem interested in me. I think I'd like you to be more interested, but somehow I feel rejection lurking in the background. I see you as not being sure how you want to relate to me.

Now write out three examples that relate to your interpersonal style and/or your experience in the group. First, identify in each of the examples above the feelings, the experiences, and the behaviors. Then make sure that your statements contain all three. Be as specific as you can. See yourself actually talking to your fellow group members or to some individual member.

Exercise 12: Concreteness and Personality Traits

This exercise should help you talk about your interpersonal style more concretely through the concretization of your personality traits.

Write down a personality trait that you believe applies to you (preferably one that affects your interpersonal style)—For example, "I am a cautious person."

Personality traits by themselves ("I am shy, powerful, caring, psychologically underdeveloped," and so on) are not very concrete. Therefore, spell out in terms of concrete experiences, behaviors, and feelings the dimensions of the personality trait you write down. For example, "I am a cautious person" may be translated into:

- I never speak my mind unless I am sure I'm right.
- I fear that others will put me down.
- I feel very guilty when I make mistakes, so I don't risk making them.
- People have hurt me in the past. They have "loved me and left me," so to speak, and I feel scarred.
- Since I have been hurt, I'm very fearful of hurting others; therefore I am reluctant to be very intimate, even with my friends.
- I have deceived myself at times. For instance, I've pretended that being self-indulgent and meeting my needs through others was caring. I think I have hurt others, and now I distrust myself and my motivation.
- I want the comfortable life more than I admit. I avoid risks that take me away from comfortable ways of living. For instance, I never give myself completely to my studies because I don't want to risk failure and don't want to surrender my need to be comfortable.
- I make few contacts with other people. I expect others to contact me.
- I allow daydreaming about interpersonal and work success to substitute for the real thing. This fulfills some of my emotional needs, and there is no risk involved in it at all.

Write down four or five of your own personality traits and then make them as concrete as possible by translating them into specific behaviors, experiences, and feelings.

CHAPTER 4: FURTHER READINGS

Watson, D. L., & Tharp, R. G. Specifying the problem. Chapter 5 in *Self-directed behavior: Self-modification for personal adjustment.* Monterey, Calif.: Brooks/Cole, 1973. Pp. 59–78.

Chapter 5

The Expression of Feeling and Emotion

As with self-disclosure, the expression of feeling, although very important in interpersonal transactions, is not an end in itself. It may be true that too many of us fear our emotions and therefore live emotionally sterile interpersonal lives. Still, the answer to this problem does not lie at the other end of the continuum, with the veritable cult of sensation and emotion that has arisen in recent years (see Back, 1972; Howard, 1970; and Lieberman, Yalom, & Miles, 1973, for commentaries on the overemphasis on feelings and emotions in human-relations training).

Just as you are continually disclosing yourself nonverbally (at least by what you do), you continually express feelings—about things and situations, about yourself, about others—through your nonverbal behavior. Ekman and Friesen (1969) refer to an inescapable nonverbal "leakage" of feelings. Therefore, since feelings are expressed even when we would prefer to hide them, some of the principles of responsible emotional expression will be discussed here in Phase I (although, of course, the expression of feeling—like self-disclosure—is a part of every phase of this model).

SOME PRINCIPLES CONCERNING THE EXPRESSION OF FEELING

You already have feelings about yourself, and you will soon develop a wide variety of feelings about your fellow group members. You may already be comfortable in expressing a wide range of emotions, or you may be somewhat fearful of such expression. What are some of the basic principles that may help you to express your feelings and emotions responsibly, both within the group and in your day-to-day interpersonal contacts? As in the case of self-disclosure, some of the following categories overlap.

Legitimacy. Feelings and emotions are part of being human. They will always seem inappropriate to you if you do not appreciate their legitimacy, if you see them as obstacles to human interchange instead of elements that add color, range, and intensity to interpersonal living. As with many other aspects of behavior, your present style of emotional expression is a *learned* style. You may well want to use your time in this lab to unlearn some of your fears about your emotions, to set aside unconstructive patterns of emotional expression, and to learn modes of expression that will enhance your interactions with others. Or you may want to learn to temper expressions of emotion that are "too much" both for yourself and for others.

Self-expression and genuineness. Let your emotions reveal who you are in the here-and-now transactions within the group. Neither manufacture emotions for effect nor disguise and repress emotions because they don't seem proper. Your expression of your feelings and emotions is part of your genuineness.

Some people have trouble expressing positive emotions. They can cry, grieve, commiserate; they can express how they feel when they are abused, hurt, and rejected; when they are down in the dumps, others know it. But they cannot express joy, affection, elation, satisfaction, peacefulness, contentment, and the like. Others can express these positive emotions, but have learned not to give expression to negative ones. Obviously, cultural norms play a large part in one's learning to hide or express emotions. Emotions in themselves, whether positive or negative, are neither good nor bad. For instance, being angry is not evil but simply human. However, when negative emotions are expressed destructively, we are faced with an entirely different question. If I express anger by striking out and injuring another, or by "holding it in," only to let it dribble out in snide remarks and obstructive behavior, then my expression of anger is not constructive. Yet many of us learn as we grow up that anger in itself (or any other negative emotion) is wrong and is to be avoided at all costs.

Constructive expression. Freedom in the expression of emotion does not mean license. Emotions can be used as weapons ("I'll dump my anger on him and that'll serve him right") or as instruments of manipulation ("If I cry, they'll feel guilty and leave me alone"). However, both positive and negative emotions can be expressed constructively. For instance, you might say to the group:

> I feel pretty cautious right now. Two people have revealed a lot of fairly sensitive stuff about themselves, and I don't think they have received much response or support.

This is an honest effort on your part to state where you are and to invite others to look into a possibly destructive situation. In this instance, mild negative emotion is expressed constructively. Compare it with the following emotional statement.

> Hell, you people are not getting anything out of me. There's no support. A guy can spill his guts out in front of you and you sit and stare at him. There's enough indifference here to choke to death on.

In this instance, emotions are "dumped" on others; such "dumping," however authentic it may seem, is usually not very constructive.

Dealing with emotions as they arise. Don't save up your emotions and then dump them on others when you can no longer hold them back. Express them when you feel them, and try to do so constructively, even though the emotions may be negative. It is much easier for your fellow group members to accept gradually increasing intensity of emotion from you than to handle an overwhelming final product all at once. Consider the following examples:

> *Marge* (in meeting number 2): Carl, I really enjoy your spontaneity. It strikes a chord in me. It's pleasant being with you.
>
> *Marge* (in meeting number 14): Carl, I've said little to you over these weeks, but I've been very much aware of you and have felt very close to you. I don't know what I can say except that I love you.

In the first statement, Marge begins to deal with her feelings as they arise, giving Carl a chance to respond. If Carl does not feel so deeply, Marge can learn this and become realistic in her expectations of response from Carl. In her second statement, however, Marge merely dumps her accumulated feelings on Carl, making it difficult, if not impossible, for Carl to respond constructively. It is too much all at once.

Sometimes we find ourselves collecting negative emotions, saving them up like trading stamps, and then cashing them in all at once by venting them on someone who is not prepared for them (see James & Jongeward, 1971, pp. 180–186, for a discussion of this "trading stamp" approach to emotional expression). You are asked to avoid this behavior in the group, and instead to deal with both positive and negative feelings while they are still manageable. This principle demands a readiness to explore negative experiences as they come up in the group. For instance, Maria might say:

> John, you've been rather quiet for a while now, and that makes me nervous. When people are overly quiet, I wonder whether I'm being observed and judged. I guess I'm wondering what keeps you from jumping into the conversation more often.

If Maria waited until the end of the group session or, worse, let a session or two go by before telling John of her feelings, it would be much more difficult for her to express her feelings constructively and much more difficult for John to respond to her creatively.

Emotional control, not repression. Emotions that are ignored or repressed tend to leak out in distorted and disguised ways. For instance, bottled-up anger "leaks out" in the form of lack of cooperation, silence, coldness, cynicism, and lightly veiled sarcasm. Such forms of emotional expression are destructive, for group members find it difficult to deal with them directly. Emotional control does have a negative side; it requires keeping emotions channeled so that they can be used to communicate rather than to punish. But, more positively, control does not require avoiding strong emotions or the strong expression of emotions. Emotional expression can be strong and disciplined. For instance,

> I'm very angry with you right now. I feel you aren't listening to me. I really feel like blowing my stack; there's something in me that wants to punish you. But that won't get either of us anywhere. I want to stay with this and face these issues that are setting us against each other. And we can get the help of other people here.

Emotional ventilation can be coupled with a desire to work through the issues that give rise to the emotion.

Emotional assertiveness. The positive side of emotional control is emotional assertiveness. Alberti and Emmons (1974) indicate that in emotionally charged situations a person can respond, generally, in one of three ways: compliantly (nonassertively), aggressively (by attacking)—and neither of these is particularly constructive—or assertively. Let's consider an example.

One of your fellow group members alternates between long periods of silence and short, very active periods of placing heavy demands on you and other group members to perform "according to the contract." When he is highly involved, he expects to be involved on his terms. In this example, suppose that he is in the middle of one of his short, active periods and is challenging you for not taking greater initiative in contacting others.

1. Nonassertive behavior. You apologize for not living up to the contract. You listen to what he has to say and respond very little. Inside you are stewing in your own juices, but you swallow your emotions. After the group meeting is over, you meet for a while with a friend of yours from the group and tell him how angry you are with this fellow. You go home feeling empty and impotent.

2. Aggressive behavior. You listen briefly to what he has to say, and then you lash out at him angrily. You accuse him of being two-faced, of not practicing what he preaches. There is no dialogue. You dump out your immediate feelings and those you have been saving for several sessions. There is some immediate relief. The topic in the group is changed, and the group settles down to deal with other issues. Afterwards, you feel some guilt and a sense of isolation.

3. Assertive behavior. You listen to what he has to say and briefly indicate that the issues he brings up are valid and that you would like to deal with them in the group. However, you also let him know that, even though the way in which he is challenging you has merit, other feelings you have are getting in the way of your dealing constructively with his challenge:

> My impression of your behavior is that for short periods of time you are totally present, and then you withdraw for relatively long periods. I've mentioned this to you before, saying how it bothers me. Now I find myself resenting your challenge, not because it doesn't have merit—it does—but because I'm not sure that we have interacted enough in this group to warrant this kind of challenge. I could accept this kind of challenge more readily from others than from you.

Assertive behavior allows for emotional ventilation, but leaves the doors of communication open.

DEGREE OF RISK IN THE EXPRESSION OF FEELING AND EMOTION

Let's start this section with an exercise.

Exercise 13: Rating Risk-Taking in the Expression of Emotion

Rate the risk taken in expressing feelings and emotions in the following cases. Rate them on a scale of 1 to 8; that is, indicate by 1 the *least* risky expression of emotion and by 8 the *most* risky expression of emotion. In each statement, imagine that two people are talking to each other.

a. ____"Last week I got very angry with you when you were poking fun at me at the dance in front of everyone. Maybe I didn't show it, but I was steaming inside."

b. ____"This morning Ken could have bitten through a nail. He was late for class and driving like mad when the bridge over the river went up and stayed up for ten minutes. Ken felt like smashing his car right into it, he was so furious."

c. ____"When Sam brought me flowers last week for my birthday, I was near tears. I'm not sure I've ever felt so warm toward anyone."

d. ____"John is really suspicious about Helen. He feels that she is manipulating him rather than dealing directly with him."

e. ____"I'm annoyed by the whole tone of the dorm regulations. Most of the rules assume that we're children in need of constant supervision."

f. ____"I care a lot about you, but I guess I'm fearful that you don't feel the same way about me."

g. ____"Peter felt that Sam, for all practical purposes, rejected him when he decided to quit the group."

h. ____"I feel relieved, because my father has relented and is willing to see me tonight."

After reading the following section, check your ratings against those in the Appendix.

What principles can be used to gauge the degree of risk involved in revealing feelings in a given situation? Johnson (1972) suggests a hierarchy of risk related to how *directly* emotions are expressed; that is, risk is related to *immediacy* of space, time, and object (person or thing).

- Are the feelings expressed toward a thing (more distant) or a person (more immediate)?
- If the feelings are about a person, is the person absent (more distant) or present (more immediate)?
- Are the feelings from the past (more distant), or are they being felt right now (more immediate)?
- Are the feelings expressed those of someone else (more distant), or are they the feelings of the speaker (more immediate)?

One possible hierarchy, then, is the following (from least to most risky):

1. Another person's feelings, past tense, about things: "Kathy was furious yesterday when she got a flat tire." The risk here is quite small because the "distance" is great.
2. *My* feelings, *present* tense, about things: "I feel bad because I haven't been able to find a suitable apartment to rent." Two elements of immediacy are added (first person, present tense), but the feelings are still about things, and therefore safe.
3. Another person's feelings, past tense, *about a third person,* both persons being absent: "Carl felt resentment toward Mary last week when she told him that she didn't want to go to the dance with him." Emotions about people are usually less safe than emotions about things. However, the other elements make this expression of feelings fairly safe.
4. Another person's feelings, *present tense,* about a third person, both persons being absent: Sue is very upset with Tom because he hasn't called at all this week.
5. *My* feelings, past tense, about a third person, who is absent: "I felt a real surge of affection for John yesterday when he encouraged me in my volunteer work." It is usually riskier to speak of your own feelings than to speak of others'.
6. My feelings, *present* tense, about a third person, who is absent: "I really love Jane, but I'm afraid to tell her so."

7. My feelings, past tense, about *you* (expressed directly to you): "I resented your silence in the group last week."
8. My feelings, *present* tense, about you and expressed directly to you: "I care about you, but your silence right now is annoying me." The risk here is the greatest because the "distance" is the least. Dealing with my feelings for you in the here and now of our interaction is called "immediacy" and will be treated more fully in Phase II.

No doubt this hierarchy could be expanded to include other elements (such as positive versus negative feelings), and there may be individual differences in its application, but the point here is not to set up absolute criteria for the difficulty of expressing feelings but to make you aware of the circumstances making emotions difficult to express. This hierarchy is complete enough to be used to score your responses to the exercise beginning this section. See page 303 in Appendix A for responses to this exercise.

THE RELATIONSHIP OF D-, M-, AND B-NEEDS TO THE EXPRESSION OF EMOTION

In the discussion of Maslow's (1968) categorization of needs in Chapter 1, I promised to attempt to tie in the various dimensions of interpersonal relating to the D-, M-, and B-needs we all have.

D-Style. The person who is still working through D-needs will tend to express himself emotionally in certain specific ways; he will probably be given to extremes of emotional expression. He may be erratic, irresponsible, uncontrolled, and unpredictable in emotional expression ("He blew up again today," "I wonder what she's going to be like today," "Boy, is he moody"), or detached, repressed, neutral, or noncommittal ("I really don't know where he stands," "I get the feeling that she's going to explode someday," "You never know what's going on inside him," "When the going got a bit rough in the group today, as usual, she withdrew"). For the D-person, emotions are fearful things or instruments of manipulation in interpersonal games or weapons in all-out interpersonal war. One way or another, emotions are the enemy.

M-Style. Emotions are no real problem for the M-person, for there is not much room for them in his rather bland approach to life. Although feelings and emotions are not actually repressed, they remain superficial and safe. The M-person seldom gets very angry, but neither is he often flooded with surprise, joy, or wonder. He lets himself become a vicarious observer of deeper emotions, which he may find in novels, TV soap operas (which he may watch avidly), or movies. Generally speaking, for him deeper feelings and emotions are things that happen to others. The M-person, both at home and in various educational systems, has been schooled in emotional neutrality and safety. Wide-ranging emotions neither disrupt his life nor add anything to it. The M-person is thus a victim of the "psychopathology of the average" (Maslow, 1968) or what Berrigan (1970) calls the "wasting disease of normalcy" (p. 58). There is not much passion in his life.

B-Style. The B-person both experiences and gives expression to a wide range of feelings and emotions. He is capable of appreciating, as well as the stronger emotions of anger, love, and awe, the more subtle emotions of curiosity, peacefulness, doubt, wonder, uneasiness, hope, and puzzlement. He expresses emotion directly, without self-consciousness or apology. He accepts feelings and emotions as a legitimate part of human living that adds color, range, intensity, and character to life. A B-person is seen not as an "emotional" person but as a sensitive person who is at ease with emotional expression and interchange. His life is not defined by emotion, but neither does he deny the importance of emotion. Feelings and emotions can be appreciated as important because they are integrated into the wider context of social-emotional living. Emotions are not pursued as ends in themselves, for the B-person has learned a lesson as old as Aristotle, who suggested that pleasure pursued for its own sake either eludes the pursuer or turns sour in his mouth when seized. Rather, emotions clarify and enhance and expand human transactions.

When you evaluate the way you integrate your emotions into your interpersonal style, you may find that your style is a mixture of the three styles outlined above. Whatever the case, the lab provides you with an opportunity to observe and reflect on the ways in which you have made emotional expression part of your interpersonal style and an opportunity to experiment with changing what you don't like.

FEELINGS DIFFICULT TO FACE

As Luft (1969) notes, there are certain emotions or emotionally colored needs that are especially difficult for most of us to face. What are some of these?

- *Feelings of inadequacy:* I'm a very ordinary person. I don't have what it takes to speak up in groups, so I always feel left out in the cold.
- *Feelings of incompetence:* Others here are so much more skillful at picking up what people say and feel. I always feel that I'm floundering.
- *Feelings of impotence:* You never seem to react to what I say here, even when it's directed to you personally. I feel washed out and useless when I'm with you.
- *Sensitivity to affection:* When you say you care for me, I tighten up completely. I can feel myself withdrawing physically.

- *Sensitivity to rejection:* You don't initiate anything directly with me in the group. I feel left out; I feel that you don't give me a chance, and that hurts.
- *Desire to punish:* When you're cynical, I get so mad I feel like calling you every name in the book. Only my pride keeps me from doing so.
- *A desire or need to be punished:* Sometimes I wish you'd haul off and hit me, at least verbally. I know you get frustrated with me, and at times I long for your attention, even if it's only your anger.
- *Guilt:* I feel despondent because I think I've let you down, betrayed you, and lied to you.
- *Depression:* I feel like I'm in a dark room. I know I'm talking to you, yet I feel absent. I've hit rock bottom.
- *Passivity:* I just sit here waiting for the rest of you to make contact with me. At the same time, I'm even afraid that you'll make contact, and that I'll have to get involved.
- *Dependency:* I get frightened when I'm away from you for any length of time.

Obviously, this list can be expanded. What feelings and emotions do you find most difficult to express?

These emotions cannot be dealt with adequately in the group unless a climate of trust and acceptance exists. Such a climate doesn't just happen; you have to work on it. This chapter suggests that there will be no climate of trust unless the members of the group are willing to take reasonable risks in revealing themselves. But this risk-taking must pay off; that is, it must be met with understanding, caring, and genuineness. The kinds of responses that engender trust and help build relationships are discussed in the next chapter.

FEELINGS ABOUT YOURSELF

Whether you reflect on them much or not, you do have a variety of feelings about yourself. These feelings change from time to time, but there is probably a certain pattern to them. It is also quite likely that these feelings—however subtle or unrecognized—influence the ways in which you interact with others. In this lab you are asked to try to get in touch with these feelings. Sometimes this is difficult, especially if you have negative feelings about yourself—feelings you have learned slowly, but very surely, from early childhood. The trouble with unrecognized negative feelings about yourself is that they have a self-perpetuating quality about them. If you feel worthless (to one degree or another), you sometimes (or frequently, as the case may be) *choose* experiences or behaviors that prove your feelings right. For instance, if, through significant interactions with others as you were growing up, you picked up the message that you were worthless (or inferior or incompetent or unattractive), and if you have unconsciously accepted that message, then you may try to experience things or do things that prove you are worthless. For instance, you are basically personable but have no friends; you are intelligent, but you fail or are a mediocre student in school; you are caring, but at the same time you alienate others by being possessive or demanding. Feelings you have about yourself *do* affect the way you interact with others and therefore are an important dimension of your interpersonal style.

On the other hand, you may well feel quite positively about yourself ("I like myself"). Or you may have some positive and some negative feelings about yourself ("Generally, I like myself, but I have some reservations about myself as a sexual person"). The point is that the lab will give you the opportunity to get in touch with your feelings about yourself, whatever they are. If you have *irrational* negative feelings about yourself (some negative feelings about yourself may be quite natural and justified; for instance, if you violate your own principles and feel guilty), then the lab may be an opportunity for you to begin a process of emotional reeducation.

EXERCISES IN THE EXPRESSION OF FEELING AND EMOTION

Exercise 14: Expanding Your Facility in Expressing Feelings and Emotions

Feelings and emotions can be expressed in a variety of ways. They can be expressed by using single words:

I feel good.
I'm angry.
I feel caught.
I feel abandoned.
I'm depressed.
I'm delighted.

Or by using phrases (idiomatic, idiosyncratic, descriptive, metaphorical):

I'm out of sorts.
I've got my back against the wall.
I'm sitting on top of the world.
I'm down in the dumps.

Or through the implications of experiential and behavioral statements:

Experiential statements (what is happening to me):
I feel like I'm being dumped on.
I feel she loves me.
I feel I'm being scrutinized, evaluated, and stereotyped.
I feel he cares.

Behavioral statements (what action I feel like taking):
I feel like giving up.
I feel like hugging you.
I feel like telling them off.
I feel like singing and dancing through the streets.

Note that feelings and emotions are expressed through implication (and thus more indirectly than directly) in experiential and behavioral statements.

I feel bad because I'm being dumped on.

"I feel bad" is, in a sense, more direct; that is, it describes the primary emotional state directly. However, experiential and behavioral statements of emotion are often more colorful and dramatic and therefore, in their own way, more direct than mere statements of primary emotions. Experiential and behavioral statements of emotion often refer, in word-economical ways, to both feeling and content. Therefore, in the statement

I feel she loves me

the implication is

I feel great (primary emotional state)
because I think she loves me (content–that is, the experience underlying the feeling).

The purpose of this exercise is to help you expand the ways in which you express feelings and emotions.

A wide variety of affective states are listed below. You are to express them in all four of the ways discussed above. In the first part of the exercise, you will be given an example in each affective category. Then you will do one of your own.

1. Joy

Single word: I'm happy.
Phrase: I'm on cloud nine.
Experiential statement: I feel he likes my work.
Behavioral statement: I feel like going out to dinner to celebrate.

Now do one of your own in the same category: joy.

Single word: ______________________________

Phrase: ______________________________

Experiential statement: ______________________________

____

Behavioral statement: ______________________________

____

2. Anger

   Single word: I'm annoyed.
   Phrase: I'm out of sorts.
   Experiential statement: I feel I'm getting a raw deal.
   Behavioral statement: I feel like telling them off.

Now do one of your own in the same category: anger.

Single word: ______________________________

Phrase: ______________________________

Experiential statement: ______________________________

____

Behavioral statement: ______________________________

____

3. Anxiety

   Single word: I'm nervous.
   Phrase: I'm on pins and needles.
   Experiential statement: I feel he's scrutinizing and judging me.
   Behavioral statement: I feel like jumping out of my skin.

Single word: ______________________________

Phrase: ______________________________

Experiential statement: ______________________________

____

Behavioral statement: ______________________________

____

4. Shame, embarrassment

   Single word: I feel naked.
   Phrase: I feel like two cents.
   Experiential statement: I feel like I've been unmasked.
   Behavioral statement: I feel like crawling under a rock.

Single word: ______________________________

Phrase: ______________________________

Experiential statement: ______________________________

____

Behavioral statement: ______________________________

____

5. Defeat

 Single word: I feel destroyed.
 Phrase: I feel done for.
 Experiential statement: I feel he's got me cornered.
 Behavioral statement: I feel like throwing in the towel.

Single word: ______________________________

Phrase: ______________________________

Experiential statement: ______________________________

____

Behavioral statement: ______________________________

____

For the following emotional states, use a single word, a phrase, and either an experiential *or* a behavioral statement.

6. Confusion

Single word: ______________________________

Phrase: ______________________________

Experiential/behavioral statement: ______________________________

____

7. Guilt, regret

Single word: ______________________________

Phrase: ______________________________

Experiential/behavioral statement: ______________________________

____

8. Rejection

Single word: ______________________________

Phrase: ______________________________

Experiential/behavioral statement: ______________________________

____

9. Depression

Single word: ______________________________

Phrase: ______________________________

Experiential/behavioral statement: ______________________________

____

10. Peace

Single word: ______________________________

Phrase: ______________________________

Experiential/behavioral statement: ______________________________

____

11. Pressure

Single word: ______________________________

Phrase: ______________________________

Experiential/behavioral statement: ______________________________

____

12. Capability, competence

Single word: ______________________________

Phrase: ______________________________

Experiential/behavioral statement: ______________________________

____

13. Low self-esteem

Single word: ______________________________

Phrase: ______________________________

Experiential/behavioral statement: ______________________________

____

14. Satisfaction

Single word: ______________________________

Phrase: ______________________________

Experiential/behavioral statement: ______________________________

____

15. Misuse, abuse

Single word: ____________________

Phrase: ____________________

Experiential/behavioral statement: ____________________

____

16. Low physical energy

Single word: ____________________

Phrase: ____________________

Experiential/behavioral statement: ____________________

____

17. Affliction, distress

Single word: ____________________

Phrase: ____________________

Experiential/behavioral statement: ____________________

____

18. Love

Single word: ____________________

Phrase: ____________________

Experiential/behavioral statement: ____________________

____

19. Constraint, hindrance

Single word: ____________________

Phrase: ____________________

Experiential/behavioral statement: ____________________

____

20. Boredom

Single word: ____________________

Phrase: ____________________

Experiential/behavioral statement: ____________________

____

21. Hope

Single word: ____________________

Phrase: ____________________

Experiential/behavioral statement: ____________________

____

Exercise 15: A Review of Feelings and Emotions

If you are to help others clarify their feelings and emotions, you should first be familiar with your own emotional states. In *How Do You Feel?*, edited by John Wood (1974), Wood and others describe in some detail their own experience of a wide variety of emotional states. These emotional states are listed below. You are asked to describe what you feel when you experience these emotions. Describe what you feel as concretely as possible: How does your body react? What happens inside you? What do you feel like doing?

Read the following examples before doing this exercise.

Example 1

Accepted: When I feel accepted:

I feel warm inside.
I feel safe.
I feel free to be myself.
I feel I can let my guard down.
I feel like sharing myself.
I feel my strengths more deeply.
Some of my fears ease away.
I feel at home.
I feel at peace.
I feel some traces of loneliness drain away.

Example 2

Scared: When I feel scared:

My mouth dries up.
My bowels become loose.
There are butterflies in my stomach.
I feel like running away.
I feel the need to talk to someone understanding.
I'm unable to concentrate.
I turn in on myself.
I feel extremely vulnerable.
Sometimes I feel like crying out.

This should not be just an intellectual exercise. Try to picture yourself in situations in which you have actually experienced these emotions, and write down what you see in your imagination.

a. The emotions: When I feel:

accepted	defensive	lonely
affectionate	disappointed	loving
afraid	free	rejected
angry	frustrated	repulsed
anxious	guilty	respect
attracted	hopeful	sad
bored	hurt	satisfied
that I belong (community)	inferior	shy
	intimate	suspicious
competitive	jealous	superior
confused	joyful	trusting

Once you have described how you feel when you experience all of these emotions, you should have a wider repertoire of words, phrases, and behavioral and experiential statements both to describe your own emotional states as they arise and to identify them in others.

b. This time, review the same emotions. Can you recall an intense or memorable experience in which each of these emotions played an important part? When did you have this experience? With whom? How intense was it?

Example

Scared:

I was going to Europe last year by plane. The captain told us that it would be a smooth flight. About an hour and a half after takeoff, it was as if we hit a brick wall. For the next three hours, we jumped and bumped and dived and swooped up. No one said anything to us. People began to get sick. I didn't realize that I could experience terror for three hours straight. I didn't talk to anyone. I was just this huge jangle of nerves for over three hours.

Now recall an experience of your own for each emotion.

c. Review the list once more. Identify the emotional areas that you find problematic. It has been hypothesized that emotions are always expressed, but that those that are difficult to express are expressed indirectly instead of directly. For instance, if I find it hard to express hurt directly, I may express it by being arrogant and controlling toward the person who has hurt me. This tendency confuses my relationship with this person. Therefore, take another look at the emotions listed on page 87. Mark those you find difficult to express. Try to indicate the indirect (and generally unproductive) ways in which you express these emotions. Study the following example.

Example

I find it difficult to express affection. I get very tense when I feel affectionate toward someone. I feel guilty about having such an emotion. I'm also afraid to express it by any physical sign, because I'm afraid of being rejected by the other. Therefore, instead of demonstrating affection directly, I do things for the other person. For instance, I give the person articles I think he will find interesting. I'm afraid that sometimes I seem to want to control the other person's life. Instead of seeming affectionate, I seem parental. This behavior backfires, for then the other person does tend to reject me.

Now do the same with emotions you find difficult to express.

Exercise 16: Feelings about Yourself

Finish the following incomplete sentences. Don't spend a great deal of time thinking about the most appropriate way to finish them; be as spontaneous as possible.

1. I get angry at myself when ____________________
2. I like myself best when ____________________
3. I feel ashamed when ____________________
4. I trust myself when ____________________
5. When I fail, I ____________________
6. I feel encouraged when ____________________
7. I puzzle myself when ____________________
8. I'm pleased with myself when ____________________
9. I get down on myself when ____________________
10. I feel confident when ____________________
11. When I violate my own principles, I ____________________
12. When I succeed, I ____________________
13. It troubles me when I ____________________
14. I'm most at peace with myself when ____________________

15. I feel good about myself when ______________________
16. When I don't understand myself, I ______________________
17. I get depressed when ______________________
18. I am buoyed up when ______________________
19. I get annoyed with myself when ______________________
20. When I take a good look at myself, I ______________________
21. When I think of what others have told me about myself, I ______________

Review the ways you have completed the sentence stems, and see if you can identify characteristic ways you feel about yourself. How do you feel about the way you feel about yourself? How do your feelings about yourself facilitate or interfere with your involvement with others?

CHAPTER 5: FURTHER READINGS

On feelings and emotions:

Egan, G. Total human expression: The elements of human dialogue—pathos, logos, poiesis. Chapter 6 in *Encounter: Group processes for interpersonal growth.* Monterey, Calif.: Brooks/Cole, 1970. Pp. 141-189.

Johnson, D. W. The verbal expression of feelings; The nonverbal expression of feelings. Chapters 5 and 6 in *Reaching out.* Englewood Cliffs, N. J.: Prentice-Hall, 1972. Pp. 85-115.

Leonard, G. B. Why we need a new sexuality. In G. Egan (Ed.), *Encounter groups: Basic readings.* Monterey, Calif.: Brooks/Cole, 1971. Pp. 260-263. (First published in *Look,* January 12, 1970.)

Levy, R. B. *I can only touch you now.* Englewood Cliffs, N.J.: Prentice-Hall, 1973.

Passons, W. R. Approaches to feelings. Chapter 9 in *Gestalt approaches in counseling.* New York: Holt, Rinehart and Winston, 1975. Pp. 183-214.

Simpson, C. K., & Hastings, W. J. Feel! Emotion workbook. Chapter 8 in *The castle of you: A personal growth workbook.* Dubuque, Iowa: Kendall/Hunt, 1974. Pp. 146-157.

Wood, J. (Ed.). *How do you feel?: A guide to your emotions.* Englewood Cliffs, N. J.: Prentice-Hall, 1974.

On assertiveness:

Alberti, R. E., & Emmons, M. L. *Your perfect right.* Revised edition. San Luis Obispo, Calif.: Impact, 1974.

Alberti, R. E., & Emmons, M. L. *Stand up, speak out, talk back.* New York: Pocket Books, 1975.

Fensterheim, H., & Baer, J. *Don't say yes when you want to say no.* New York: Dell, 1975.

Lazarus, A., & Fay, A. *I can if I want to.* New York: William Morrow, 1975.

Liberman, R. P., King, L. W., De Risi, W. J., & McCann, M. *Personal effectiveness: Guiding people to assert themselves and improve their social skills.* Champaign, Ill.: Research Press, 1975.

Smith, M. J. *When I say no, I feel guilty.* New York: Dial Press, 1975.

PART 3

Phase I: The Skills of Listening and Responding

So far, we have examined the communication process from the point of view of one person trying to communicate ideas or emotions to another. We have seen that is impossible for us *not* to reveal ourselves, for what we say and what we do constitute a continuous source of self-revelation. Communication, however, is obviously two-sided. If I reveal myself to you in some way, I usually expect you to respond to me. If we are to involve ourselves in each other's lives to one degree or another, there must be a *dialogue* between us; our interactions must be marked by a great deal of *mutuality.* Just as there are skills of self-disclosure, there are skills of response. If I reveal myself to you, what (whether I reflect on it or not) do I expect of you? Generally speaking, understanding. All of us have a deep desire to be understood; this desire is an important dimension of our humanity. This does not mean that the communication of understanding is the only response that can be given to self-revelation. If you reveal yourself to me, I might also reveal myself to you, or probe more deeply into what you have said, or challenge either what you say or some of its implications, or start helping you solve some problems. But these later forms of communication will ordinarily make more sense if I first try to understand what you have said. The communication of understanding is a kind of oil that lubricates the entire communication process and makes all kinds of interchanges go more smoothly. More on understanding in Chapter 6.

First we must backtrack a bit, for there is something I must do—and do well—*before* I communicate understanding to you. I must first attend to you and listen carefully to you. Poor attending and poor listening will result in poor understanding. And without good attending, listening, and understanding, we can expect little mutuality and little benefit from our other forms of interaction. In a very real sense, interpersonal communication stands or falls on the skills of good responding. Therefore we begin with the skills of attending and listening, for without these skills adequate responding is simply impossible.

One caution: Brammer (1973), discussing counseling or helping skills, notes that "helping relationships have much in common with friendships, family interactions, and pastoral contacts. They are all aimed toward fulfilling basic human needs, and when reduced to their basic components, look much alike" (p. 48). The "basic components" he mentions are, principally, the skills we are about to discuss—both the skills of responding and the skills of challenge. These skills belong first and foremost in everyday life, and for this very reason are also, often in an intensified form, the core skills of the helping process. Thus Carkhuff and Berenson (1967) can talk about counseling as a "way of life," for the skills they discuss are the skills needed in marriage, friendship, family living, and the interpersonal aspects of work situations.

The purpose of your training in the skills of responding and the skills of challenge is *not* to turn you into counselors or helpers in your interpersonal relationships. In fact, if you present yourself primarily as a helper in your training group, you will probably be rejected. Helping is a one-way relationship; it implies that one person is functioning well whereas the other is not. The skills to which you are about to be introduced are skills in the service of *mutuality*, not of helping. Friendship is a two-way relationship. Friends do help each other, and they do use the skills of responding and the skills of challenge to do so. However, this helping is a by-product of their mutuality, and neither friend takes on the role of helper or helpee. The best friendships involve both support and challenge. The skills illustrated in the following chapters are the "guts" of both effective support and effective challenge.

Chapter 6

Attending and Listening

SOCIAL INTELLIGENCE: THE DISCRIMINATION/COMMUNICATION DISTINCTION

A good communicator is a perceptive communicator. He attends carefully to the other person and listens to both his verbal and his nonverbal messages. He clarifies the messages through his interaction with the other and builds further communication patterns on what he hears and understands. He is in touch with his own thoughts and feelings and with how they interact with those of the other. In sum, a good communicator is socially intelligent.

The history of scientific inquiry into the definition and measurement of social intelligence need not concern us here (see Walker & Foley, 1973), but the basic division of social intelligence—into (1) the ability to understand social situations (for example, the kinds of individual and group communications discussed in this book) and (2) the ability to act on this understanding—is an important one: The discrimination/communication distinction. Carkhuff (1969) calls the ability to understand the various components of a social or communication situation "discrimination"; Gazda (1973) calls it "the act of perceiving." In the introduction to Part 2, I discussed the "awareness" component of interpersonal skills. You can be a good discriminator, or perceiver, in a number of ways:

- As you interact with another, you understand what is happening inside yourself (for example, you are in touch with your feelings of frustration or hostility or affection as you listen to someone in your group).
- You perceive what is happening in your environment (for example, in your group you see that Bill and Nancy don't communicate well because neither takes time to understand the other).

- You perceive what is happening in the "world" of the other person; you understand the other person's frame of reference and not just your own (for example, in your group you get along well with Paul, but you also see that Jim does not, and that Jim resents Paul's competence in communication because it is a competence he doesn't possess).
- You see ways in which members of your group are communicating well; you also see patterns of communication that are self-defeating (for example, you see that Chris allows himself to get bored with group interaction, and that eventually he gets frustrated and angry and then "dumps" his emotionality on the other group members in a way that is self-defeating, since he merely arouses their resentment).
- You know what challenges may help others (for instance, you see that it may be helpful to point out to Mary that you think she plays the game of "poor little girl" in her interactions with men in the group).
- You see what kinds of programs may help others (for example, you see that the quality of Terry's communication would be improved by his learning to be more assertive in his interactions with others).

All of these skills are related to *perception;* but perception is not enough for good communication, which is the ability to *act* in a facilitative way in social situations (Carkhuff, 1969). A good communicator can translate his or her perceptions, insights, and discriminations into effective interpersonal transactions. He or she is not just a perceiver but also a doer. I use the term "agent" (the opposite of "patient") to refer to a person who exercises initiative in his own physical, intellectual, and social-emotional growth and in the process of communicating with others (see Egan, 1970, pp. 359–362). If you are a good communicator, you will exercise "agency" in a number of ways:

- Your being in touch with what is happening within you will enable you to reveal yourself to others appropriately and thus develop mutuality (for example, you can share with Jim your own struggles with feelings of incompetence in relating to others).
- Your understanding of what is happening in the environment can be shared (for example, you share with your fellow group members your perception of the group climate as an overly cautious one; you share your feelings that the group has implicitly established certain norms that forbid the discussion of certain topics).
- You translate your perception of what is happening in the "world" of the other into the communication of accurate empathic understanding (for instance, you communicate to Jim understanding of his feelings of resentment toward Paul and frustration with his own inability to communicate more effectively).
- You translate your perception of the need for challenge into caring and responsible challenges based on understanding (for example, you invite Mary to examine the way she relates to the men in the group and to get feedback from other group members.)

In a word, the good communicator is an *agent* rather than an observer, however perceptive, in his interactions with others. That a person is a good discriminator does not mean that he is automatically a good communicator. On the other hand, a good communicator must be a good discriminator, for a good communicator is one who acts on *accurate* discriminations. Let's consider an example.

> *Group Member A:* I'm not sure I should be in this group. All of you seem to be doing much better than I. I wonder whether you just got a head start or I'm that far behind. What comes easy to you just doesn't come easy for me. I'm not so perceptive, I'm not so articulate. I seem to get in the way rather than contribute.

A good discriminator will see that this person is depressed, that he feels frustrated, that he is down on himself, that comparing himself with others is getting him nowhere, that he feels incompetent, and that he seems ready to throw in the towel. But unless you can communicate the substance of your observations to him, you remain merely a good discriminator, and discrimination is not enough to help the communication process. However, if you are a good communicator, you might say:

> You just don't feel so competent as the other members of the group. You're depressed, and maybe you even want to throw in the towel.

The good communicator does not allow his understanding or perception to remain locked up inside himself.

The following pages will be filled with many examples to help you learn how to distinguish good communication from bad. However, the only way to learn effective communication is by communicating. Therefore, you must practice. There is no substitute for doing.

ATTENDING AS PHYSICAL AND PSYCHOLOGICAL PRESENCE

Deep interpersonal transactions demand a certain intensity of presence. But this presence, this "being with" another person, is impossible without attending. Attending seems so simple a concept to grasp and so easy to do that you might ask why so many words are being devoted to it. As simple as attending is, it is amazing how often people fail to attend to one another. This is certainly true in human-relations-training groups. You have probably heard one person say to another (or you have said it, or it has even been said to you): "You're not listening to what I'm saying!" Or someone reads a magazine while you are talking to him, or it becomes obvious that the person at the other end of the telephone conversation is eating lunch, reading, or engaging in some activity that prevents him from giving his complete attention to you. When a person is accused of not attending, his answer is predictable: "I can repeat word for word everything you've said." Since such a reply brings little comfort to the accuser, attending must certainly be more than the ability to repeat someone's words. You don't want the other person's ability to remember what you have said. You want *him*. You want more than physical presence; you want psychological presence.

An engaging example of the impact of attending behavior is given by Ivey and Hinkle (1970; also see Ivey, 1971, pp. 93–94). At a prearranged signal, six students in a psychology seminar switched from the traditional student's slouched posture of passive listening and note-taking to attentive posture and active eye contact with the teacher. During the nonattending condition, the teacher lectured from his notes in a monotone, using no gestures, and paying little or no attention to the students. However, once the students began to attend, the teacher began to gesture, his verbal rate increased, and a lively classroom session was born. At another prearranged signal later in the class, the students stopped attending and returned to typical passive student posture and participation. The teacher, "after some painful seeking for continued reinforcement," returned to the unengaging teacher behavior with which he had begun the class. Both students and teacher got what they deserved: reciprocated inattention. But simple attending changed the whole picture.

Attention is a potent reinforcer. Erikson (1964) speaks of the effects of both nonattention and negative attention on the child:

> Hardly has one learned to recognize the familiar face (the original harbor of trust) when he becomes also frightfully aware of the unfamiliar, the strange face, the unresponsive, the averted. . . . And here begins . . . that inexplicable tendency on one man's part to turn elsewhere [Erikson, 1964, p. 102].

Perhaps the averted face is too often a sign of the averted heart. At any rate, most of us are very sensitive to others' attention (or inattention) to us. In light of this fact, it is curious how insensitive we can be about attending to others. Since attending is reinforcing, it is an important element in the social-influence dimension of the communication process. If you attend carefully to another person, you are more likely to make an impact on him than if you do not. Conversely, you will ordinarily be more open to influence from those who attend carefully to you. Life can be fuller if I have an impact on others and allow others to have an impact on me (provided that having impact is not synonymous with manipulating). Attending, then, is associated with power (in a good sense of the word) in interpersonal relating.

Physical Attending

Your body plays a large part in verbal communication. Therefore, attending requires becoming aware of how you use your body in the communication process. You should adopt a posture of involvement in your interactions with others. The basic elements of physical attending can be recalled through the help of the acronym SOLER:

- S—face the other person *squarely*. This is the basic posture of involvement. It says "I am available to you." Turning at an angle from another person lessens your involvement.
- O—adopt an *open* posture. Crossed arms and crossed legs are often at least minimal signs of lessened involvement. An open posture is a sign that you are open to what the other person has to say and are open to communicating directly with the other. It is a nondefensive position.
- L—*lean* toward the other. This is another sign of availability, presence, or involvement. In your everyday environment, when you see people who are seriously involved in a conversation, you see them leaning forward as a natural sign of their involvement.
- E—maintain good *eye* contact. As you speak with another person, you should spend much of the time looking directly at him. Some balk at this, saying that the other person will feel "stared down." However, there is a difference between staring a person down (which is a power function or perhaps a sign of rigidity) and the kind of eye contact that is both indicative of and facilitative of deeper involvement. Watch two people intensely involved in a conversation with each other. Their eye contact is almost uninterrupted. They are not self-conscious, for the *interaction* is so important and so engaging that they do not reflect on just how much time they spend looking directly at each other.
- R—be at home and relatively *relaxed* in this position. Relative relaxation says to the other person "I'm at home with you." The physical position described here has a kind of tension about it. However, an effective communicator *is* relatively comfortable with involvement and intimacy, and therefore will be relatively relaxed even in this attending position. Trying to seem less relaxed than you really are is not very genuine, but intensity of interaction eventually will not be a chore for you but will come naturally, enabling you to be at home with intensity.

In no way do I suggest here that you lock yourself rigidly into the attending position described. Attending should serve the communication process; it is not an end in itself. The important thing to learn is that your body *does* communicate—for good or ill. What you do with your body does punctuate and modify your verbal messages. The high-level communicator does not adopt any rigid formula for attending. He flows naturally with the communication process and adopts whatever posture is called for at any given moment during the interaction. But, at least subconsciously, he is aware that his body communicates, and he *uses* his body to communicate. If you are comfortable in communicating with others, your body movements will not be rigid and you will not be afraid to use your body to communicate. Strong, Taylor, Bratton, and Loper (1971), in a study of nonverbal counselor behavior, discovered that the counselor who remains too still (perhaps rigid) is seen as precise, reserved, thoughtful, serious, orderly, controlled, cold, aloof, and intellectual. In contrast, the counselor who is physically active (although not fitful) is seen as friendly, warm, casual, and carefree (relaxed and role-free), and therefore more attractive. I suggest that this comparison could be made among communicators in general.

I once videotaped one of the last sessions of a human-relations-training group. One of the participants was slouched down in his chair, his legs stretched out in front of him. Another member asked him how he had liked the group experience. He seemed to be caught somewhat unaware and answered, hesitatingly, that he had "really enjoyed" the sessions. During the replay, when this part of the tape was played, he cried out "Turn it off! What a liar I am!" His whole body had cried out the real message. Good physical attending is essential for good group interaction. Even if only one or two members of an eight-member group drop out of the picture physically through nonattending, much of the immediacy of the interaction is lost.

Listening as Psychological Attending

Physical attending is a manner of being present to another; listening is what you do while you attend. You listen to both verbal and nonverbal behavior. We are only beginning to realize the importance of nonverbal behavior and to study it scientifically. Take the case of the ordinary smile:

> To illustrate the complex, intricate nature of this medium [facial expressions], let us consider a very simple form of facial expression—an ordinary smile. One of the most common instances is the "simple smile," a mere upward and outward movement at the corners of the mouth. It indicates inner bemusement; no other person is involved. The "upper smile" is a slightly more gregarious gesture in which the upper teeth are exposed. It is usually displayed in social situations, such as when friends greet each other. Perhaps the most engaging of all is the "broad smile." The mouth is completely open; both upper and lower teeth are visible.
>
> Yet without other facial movements, particularly around the eyes, smiles would not really mean what they seem to mean. For appropriate warmth, the upper smile is usually enhanced by slight changes around the outer corners of the eyes. Even the broad smile is not always an entirely convincing expression of surprise or pleasure unless it is accompanied by an elevation of the eyebrows. Other emotional expressions also depend upon a delicate use of the eye area. In a sad frown, the eyebrows will ordinarily be drawn down at the outer ends. By contrast, they will be depressed on the inside in an angry frown [Saral, 1972, p. 474].

The face and the body are extremely communicative. Even when two people are silent together, the atmosphere can be filled with "messages."

Mehrabian (1971) reports on research that he and his associates have done in the area of nonverbal messages and inconsistent messages, such as liking/disliking.

> One interesting question now arises: Is there a systematic and coherent approach to resolving the general meaning or impact of an inconsistent message? Indeed there is. Our experimental results show:

> Total liking equals 7% verbal liking
> plus 38% vocal liking
> plus 55% facial liking
>
> Thus the impact of facial expression is greatest, then the impact of the tone of voice (or vocal expression), and finally that of the words. If the facial expression is inconsistent with the words, the degree of liking conveyed by the facial expression will dominate and determine the impact of the total message [Mehrabian, 1971, p. 43].

Both nonverbal behavior (such as bodily movements, gestures, and facial expressions) and paralinguistic behavior (such as tone of voice, inflection, spacing of words, emphasis, and pauses) should be listened to, for you must respond to the other's total message and not merely to his words. Indeed, as Mehrabian's research illustrates, nonverbal and paralinguistic cues can contradict the overt meaning of words. For instance, tone of voice can indicate that a verbal "no" is really a "yes." The high-level communicator, since he listens to all cues and messages, is ready to respond to the *total* communication of the other. Although only a few basic dimensions of human feelings and attitudes are conveyed nonverbally—like, dislike, potency, status, responsiveness—they are extremely important qualifiers of verbal messages.

Nonverbal behaviors are often carriers of the emotional dimensions of messages. For example, the person speaking kicks his feet, maintains a half-smile, wrings his hands, grimaces, or folds his arms over his chest. As such, nonverbal behaviors constitute a more primitive and less easily controlled communication system than verbal behaviors. How often, then, do you know what they mean? The high-level communicator listens to the entire context and does not simply fix his attention on details of behavior (for instance, he places what a group member is saying in the wider context of what is—or is not—happening in the group at that particular time). Nonverbal behavior helps punctuate and color the interaction; it confirms or denies the message carried by the words; it adds feelings not contained in the words themselves. The high-level communicator is aware of and uses the nonverbal communication system, but he is not seduced or overwhelmed by it.

Some lab participants, in the early stages of the training process, hesitate to label another's feelings in responding to him unless the speaker states them explicitly. However, the speaker can show his feelings (for example, anger) by both content and manner of expression without saying explicitly "I am angry." Feelings can be carried by verbal content and tone of voice and by all the elements of nonverbal behavior. Many feelings expressed this way are what Gazda (1973) calls "surface" feelings (p. 68), and it is perfectly legitimate for you to communicate understanding of these feelings to a fellow group member. However, if feelings are indistinct or too deeply buried or camouflaged, you should be hesitant to attribute them to the speaker. High-level attending or "being with" another person is absolutely necessary for high-level discrimination.

Obviously, you also listen to verbal messages. In the total communication of the speaker, you listen primarily for *content*—that is, the experiences and behaviors of the person speaking (if he is revealing himself) and the feelings associated with them. Good listening supplies you with the building blocks of good understanding. The defensive person finds it difficult to listen, for he is afraid of hearing things he doesn't want to hear. He develops the skill of selective listening in order to keep his anxiety in line. For instance, he hears another person when the person talks superficially but not when he talks intimately and personally. He hears praise but not criticism (or vice versa). He hears parts of sentences, but disturbing words are left out. Obviously, the good communicator cannot be a selective listener.

A caution: the process of human communication is presented in this book in bits and pieces. The low-level communicator often fastens selectively on this bit or that piece. For example, he will become intrigued with nonverbal behavior and will make too much of a half-smile on the face of another person. He will seize the smile and lose the person. The high-level communicator, on the other hand, can integrate the pieces. He makes communication skills—from attending on—his own and doesn't become their victim.

Exercise 17: Body Language

The purpose of this exercise is to acquaint you with a sampling of forms of nonverbal behaviors (body language) that carry messages. The behaviors listed below might accompany the words of a speaker, or they might be the behaviors of group members who are silent at the moment.

You are in a group meeting, observing your fellow group members. You observe, at one time or another, the following nonverbal or paralinguistic (voice-related) behaviors. Without any further context, try to indicate what possible meanings each of these behaviors can have. Try to find more than one meaning for each behavior.

1. A person nods his head up and down
2. A person turns her head rapidly in a certain direction
3. A person smiles slightly
4. A person's lower lip is quivering
5. A person speaks in a high-pitched voice
6. A person's voice is monotonous
7. A person opens his eyes wide suddenly
8. A person keeps her eyes lowered as she speaks to someone else
9. A person's nostrils are flaring
10. A person raises one eyebrow
11. A person shrugs his shoulders
12. A person has her shoulders forced back
13. A person has his arms folded tightly across his chest
14. A person is waving her arms
15. A person is holding his chair tightly with his hands

16. A person hides part of her body with her hands
17. A person's breathing is quite irregular
18. A person inhales quickly
19. A person digs his heels into the floor
20. A person continuously moves her legs back and forth
21. A person sits with her arms and legs folded tightly
22. A person is constantly squirming
23. A person holds her body bent to one side
24. A person has a tic
25. A person is getting pale

Compare your meanings with those of a partner, or share them with all of the members of the group. How many different meanings are associated with each kind of behavior? What are the implications of this fact?

Notice that most of these behaviors have more than one possible interpretation. For this reason, both verbal and nonverbal behavior must be interpreted in the light of the total communication context—(1) the verbal and nonverbal behavior of the speaker and (2) the environment in which his communication is taking place.

Exercise 18: The Importance of Nonverbal Cues for Verbal Communication

The purpose of this exercise is to give you some idea of how much you depend on nonverbal cues in the interpretation of verbal messages.

Choose a partner from among your fellow group members. Sit facing your partner. *Close your eyes.* Have a two- or three-minute conversation on some topic related to the goals of this laboratory experience; keep your eyes closed during the entire conversation.

Share with each other what you experienced during the conversation. What nonverbal cues did you miss the most? In what ways was the conversation stilted?

Exercise 19: Attending and Recall

This exercise is meant to help you sharpen your attending skills with respect to the nonverbal dimensions of the communication process.

Choose a partner from among your fellow group members. Spend two minutes in silence just looking at this person—the features of his face, his body build, the arrangement of his limbs—all that pertains to his body. After two minutes, close your eyes and describe this person's physical appearance to him from memory. If you cannot give a complete description after a two-minute survey, open your eyes, take in the details you have missed, close your eyes once more, and then complete the description. After you've finished, let your partner go through the same process.

Exercise 20: Experiencing Nonattending in a Group Conversation

1. Have all of the members of the training group (six to eight members) meet in a circle for a group discussion.
2. Choose a topic related to training (for example, what fears you have as you move into the training experience).
3. Have a discussion on the topic for three or four minutes, during which half of the members attend while the other half assume various nonattending positions.
4. After three or four minutes, the nonattenders now attend, while the attenders now assume various nonattending positions.
5. Stop the discussion and discuss the effects of attending and nonattending as concretely as possible.

Exercise 21: Degrees of Attending in a One-to-One Conversation

1. Choose a partner from among the members of your training group. Your partner will be the "speaker" and you will be the "responder."
2. Choose a topic of conversation related to training.
3. Start the conversation in a minimally attending position.
4. At a signal from the trainer, move into an intensely attending position. Continue for another two or three minutes.
5. Stop the conversation and discuss the different effects of different degrees of attending, from the viewpoints of both the one communicating and the one being communicated to.

Exercise 22: Degrees of Attending in a Group Conversation

1. Have all of the members of the training group meet in a circle for a group discussion.
2. Continue discussing the topic from Exercise 21 above.
3. In the beginning, the members should assume only minimally attending positions, but all should attend.
4. After a few minutes of discussion, at a signal from the trainer, all of the members should move to intense attending positions. Continue the conversation for another few minutes.
5. Stop the conversation and discuss the differences felt between minimal and intense attending positions.

Exercise 23: Listening to Nonverbal and Paralinguistic Cues

Physical attending has two functions: (1) it is a sign to the other that I am actively present and working with him, and (2) it helps me to be an active listener (that is, my physical attending helps my psychological attending).

What do I listen to? I listen to both the verbal and the nonverbal messages of the person speaking. The purpose of this exercise is to become more sensitive to nonverbal and paralinguistic cues and messages. (Remember, "paralinguistic" refers to the way one uses one's voice in the communication process; paralinguistic cues include tone of voice, loudness, pitch, pacing of words, stumbling over words, grunts, sighs, and so on.) Nonverbal and paralinguistic cues have two general functions: they may (1) confirm, punctuate, emphasize, modulate, or otherwise modify the verbal messages of the speaker; or sometimes, they may (2) contradict the verbal message of the speaker and thus contain the real message. For instance, if the speaker raises his voice and pounds on the table while delivering an angry message, then both raising the voice (a paralinguistic cue) and pounding on the table (a nonverbal behavior) underscore and emphasize his anger. However, if the speaker says in a very hesitating way (a paralinguistic cue) while fidgeting with his hands (a nonverbal cue) that yes, he would like to go out to dinner that night, his nonverbal and paralinguistic cues contain the real message —one that contradicts the verbal message.

Directions

Trainees should divide into groups of four (Members A, B, C, and D). Members A and B should spend five or six minutes discussing what they like and don't like about their present interpersonal styles (or any other topic relevant to the training experience). Members C and D act as observers. While A and B are speaking, C and D should take written notes of A's and B's nonverbal and paralinguistic behavior. The speakers' behavior should be observed and noted, but C and D should take care not to overinterpret this behavior. After five or six minutes, C and D should give A and B feedback on the highlights of their nonverbal and paralinguistic behavior. Then roles are switched and the process repeated, with A and B becoming the observers and C and D the interactants.

Samples of Typical Feedback Statements

- Most of the time you spoke very quickly, in spurts. It gave me a feeling of tension or nervousness.
- You sat very still throughout the dialogue. Your hands remained folded in your lap the whole time, and there was practically no bodily movement. The position made you look very "proper," and it gave me an impression of shyness or rigidity.
- When you talked about being a very sensitive person, one who is easily hurt, you began to stumble over your words a bit. The message seemed to be that you are sensitive about being so sensitive.
- You tapped your left foot almost constantly.
- You put your hand to your mouth a great deal. It gave me the impression of hesitancy on your part.
- When B began to talk hesitantly about being shy, you leaned back and even moved

your chair back a bit. I'm not sure whether you were showing him that he made you uncomfortable, or whether you were easing off, giving him room to speak.

- You broke eye contact a great deal when you were talking about yourself, but not when you were listening.
- You were so relaxed—at times you even slouched a bit—that you almost gave me the impression that you were uninterested in the whole task.

A Caution

The purpose of this exercise is to make you aware of nonverbal and paralinguistic cues. At this stage, you should begin to note the amount and variety of such behavior, but be slow to interpret such behavior. As you gradually become more aware of nonverbal and paralinguistic behavior, your abilities to interpret and read these cues intelligently will grow along with your experience. However, great care should be taken to interpret such behavior only within the *total* communication context (which includes who is talking to whom, under what circumstances, with what antecedents, on what topics, and so on). For instance, a drooping head and a stony expression can mean dejection or anger or despair, but such behavior does not invariably indicate any one of these emotions. Finally, if you become overly preoccupied with nonverbal and paralinguistic cues and their interpretation, you will become overly self-conscious about the communication process itself, and such self-consciousness is self-defeating.

Exercise 24: "Right Now I Am Aware . . . ."

The purpose of this exercise is to help you focus on the nonverbal behavior of a partner and on your own feelings, experiences, and behaviors (both overt and covert) as you interact with another person.

Directions

Choose a partner from among your fellow group members. Sit facing each other. Take turns saying single sentences to each other, each sentence beginning with the stem "Right now I am aware. . . ." Finish the stem by making some observation about the nonverbal behavior of the other or about your own feelings, experiences, or behaviors (overt or covert). Spend five minutes or so in this dialogue.

An Example

A: Right now I am aware that you are leaning toward me.
B: Right now I am aware that you have a slight smile on your face.
A: Right now I am aware that I am grasping the side of my chair fairly tightly.
B: Right now I am aware that my mouth is a bit dry.
A: Right now I am aware that I'm glad that it's you who are my partner.

B: Right now I am aware that I feel caught off guard by your last statement.
A: Right now I am aware of a kind of tightening in my stomach.
B: Right now I am aware that this is the first time we have been in such close contact.
A: Right now I am aware that your body, or at least your face, is a bit more relaxed.
B: Right now I am aware that we are talking much more personally than I had anticipated.

How does it feel to be forced into a kind of micro-attending, focusing both on yourself and on the behavior of your partner? What kind of contact did the two of you make? What did you learn?

This exercise forces you to look at a number of the details of your interactions. It can sensitize you to important cues within yourself and in the behavior of others—cues that signal what is happening in the communication process. At the same time, remember that there is danger in becoming overly sensitive to details, becoming too much of a miniaturist in your approach to communication. You should maintain a balance between the larger communication picture and its component parts.

Exercise 25: Identifying Feelings

The next two exercises give you practice in *discrimination* (perception) and therefore will help prepare you for learning the skill of accurate empathy, which involves direct *communication* to another.

Directions

Read the following statements and write down a number of adjectives or phrases that accurately identify what the speaker is *feeling*. Remember from the last chapter the number of different ways in which emotions can be expressed or identified. In the next exercise you will be asked to deal with *content*.

The following "group member" statements are made up, not taken from actual dialogues. The statements are, for the most part, direct, concise, and intense. As such, they represent an ideal in interpersonal communication. If you cannot yet speak with the same directness, conciseness, or intensity, these examples are not meant to discourage you but rather to stimulate you gradually to learn the same kinds of skills. Read each statement and ask yourself: how does this person feel?

Sample Statements

1. John, I'm finding it hard to relate to you. It seems as if you talk down to me because I'm a woman. I feel as if I have to prove myself to you. Part of this reaction may be coming from the way my relationships with men have been

going lately. I'm determined not to be a second-class citizen, yet with you I somehow feel forced into that role.

2. Joan, we've been together in this group for eight weeks now, and I think I've shared quite a bit of myself with you and the others here. I've really felt your caring and warmth in response to me. At the same time, though, I feel a lack in our relationship. I don't really feel I know you, other than that you respond accurately and with warmth. There's a whole other side missing.
3. Bill, it's good you're back today, because I'm not sure where we stand with each other. Two weeks ago, as we were leaving, I told you I was sorry you'd have to miss last week's class. You kind of brushed off what I said—almost as if you didn't believe I really cared whether you were here or not.
4. Jean, I appreciate your feedback about my style of confronting. I know I come on too strong for some. I find it hard to temper my feelings. I'd like to be able to work on that in this group, yet I'm a little afraid that I might alienate you if I'm too direct with you.
5. Boy! These silences! When I was a kid the family used to eat meals in silence, and I get the same feelings when we have silences here. I mean, don't we have anything to say? My bet is that we have too much to say!
6. Dottie, for weeks now I've felt that something hasn't been quite right between us. I think I've finally put my finger on it. You're very much like a woman who was in charge of me when I was in boarding school. I had a really hard time getting along with her, and I think I'm focusing some of the resentment I felt toward her on you. I thought for a while that it was just you, but now I realize that my feelings from the past are blocking you out.
7. Frank, I've been thinking about you this past week, and—well, I don't know what to say. I really like your sensitivity and directness. I see you as a really caring person, but you always seem to turn to Mary for response or support, and I end up feeling that I don't know where I stand with you.
8. Sally, you and I seem to be able to have a lot of fun together quite easily. I've noticed, though, that whenever you try to get serious I feel myself pulling away. It's almost as though I don't want you to come into my life any further than you have—and when I look at it honestly, that's not really very far.
9. Now we're getting somewhere! You've finally heard that I'm not a little child anymore, that I don't need another set of parents here. You two have been "helping" me all along, and I've been the nice boy accepting your help. Getting rid of the game should do us all some good!

Compare your responses with the suggested ones in the Appendix and with those of your fellow trainees. Pick out the words or phrases that fit best for you.

Exercise 26: Identifying Content (Experiences/Behaviors)

This exercise is meant to help you focus on content. Once you have identified feelings, the next step is to identify what concrete experiences and behaviors underlie these feelings.

Reread the speaker statements in the previous exercise; then complete the following statements (here the feelings have already been identified) by identifying the experiences and/or behaviors underlying the feelings expressed.

1. She feels resentful because
2. She feels dissatisfied because
3. He feels rejected and hurt because
4. He feels hesitant and apprehensive because
5. He feels anxious because
6. She feels relieved because
7. She feels resentful and left out because
8. She feels uncertain and a bit distressed because
9. He feels great because

Compare your responses with those suggested for this exercise in the back of the book. Also compare these responses with those of your fellow trainees. Which of the lot seem best to you?

CHAPTER 6: FURTHER READINGS

On attending:

Carkhuff, R. R. *The art of helping*. Amherst, Mass.: Human Resource Development Press, 1972. Pp. 12–45.

On nonverbal behavior:

Knapp, M. L. *Nonverbal communication in human interaction*. New York: Holt, Rinehart and Winston, 1972.

Mehrabian, A. *Silent messages*. Belmont, Calif.: Wadsworth, 1971.

Passons, W. R. Nonverbal awareness. Chapter 6 in *Gestalt approaches to counseling*. New York: Holt, Rinehart and Winston, 1975. Pp. 101–130.

Chapter 7

The Communication of Accurate Empathic Understanding: Creating a Climate of Support

The rest of good listening is good responding. When a person says "I am a good listener," he sometimes means that he attends well to what the other person is saying, that he gives the person the opportunity to express himself; but he doesn't always mean that he responds well, that he *dialogues* well with the other person. Listening without response is usually barren.

THE COMMUNICATION OF PRIMARY-LEVEL ACCURATE EMPATHIC UNDERSTANDING (AE I)

There are two levels of accurate empathy (see Carkhuff, 1969), which I call primary-level accurate empathy and advanced-level accurate empathy (AE I and AE II). But before we consider the differences, let's take a brief look at the communication of accurate empathic understanding in general.

You are accurately empathic if you (1) discriminate—get inside the other person, look at the world from his perspective or frame of reference, and get a feeling for what his world is like; and (2) communicate to the other this understanding in a way that shows him that you have picked up, generally, both his feelings and the experiences and/or behaviors underlying these feelings. As Mayeroff (1971) puts it:

> To care for another person, I must be able to understand him and his world as if I were inside it. I must be able to see, as it were, with his eyes what his world is like to him and how he sees himself. Instead of merely looking at him in a detached way from outside, as if he were a specimen, I must be able to be *with* him in his world, "going" into his world in order to sense from "inside" what life is like for him, what he is striving to be and what he requires to grow [Mayeroff, 1971, pp. 41-42].

If one of your fellow group members is looking at the floor, hunched over, and says something like:

> I'm not revealing myself, not getting myself over to you people. I didn't know that I would be this wrapped up in myself. Everything I say seems so superficial, and yet I see a lot of you really putting yourselves on the line. Yet this is the way I am outside the group, too—with my family, with my friends. There's just not a lot of contact between the person I think is me and others.

You might begin to respond to him by saying:

> You sound really frustrated and depressed. It's hard for you to let others inside to find out who you are. And being in this group seems to make you more aware of it than ever.

You see his frustration and depression (*feelings*) and begin to understand what underlies these feelings—his inability to act effectively in interpersonal situations (*content*)—and you communicate to him this understanding of his world. This is accurate empathy. Or, a friend might tell you that she has just graduated as a nurse and has been accepted by the hospital where she has always wanted to work—all a dream she has had since she was a little girl. You might say something like:

> I've never seen you bursting with joy like this! It must feel great to finally accomplish what you've set out to do.

This, too, is accurate empathy. One thesis of this book is that the importance of accurate empathy in human communication can hardly be overestimated, for it has significant pragmatic value in all the deeper interactions of life.

Primary-Level Accurate Empathy (AE I)

Primary-level accurate empathy is a communication to the other person that you understand what he says *explicitly* about himself. In AE I, you don't try to dig down into what the other person is only half-saying, or implying, or stating implicitly. You don't try to interpret what he is saying, but you do try to get inside his skin and get in touch with *his* experiencing. Let's take a look at a few examples of AE I.

> *Group Member A:* I've had other group experiences, and nothing has ever really happened. I don't even know why I'm trying again. Yet I'm still dissatisfied with my relationships with others. I guess that's why I'm here.
>
> *Group Member B:* You're depressed because you're not sure that this group experience is going to work any better than the others, but you feel you have to keep trying.

Group Member C: When it comes right down to it, I'm not very generous at all. I'm very choosy about the people with whom I'm going to relate. As I look at my relationships, most of them—there aren't that many—are on my terms. It's almost as if anyone who is going to relate to me is going to have to do it my way. Hmm. I've never said that out loud before.

Group Member D: You pretty much control what's going on in your interpersonal life—and more or less make sure it goes your way. Just saying it seems to make it dawn on you in a new way. You sound disappointed in yourself.

Group Member E: I did that exercise on interpersonal strengths and weaknesses last night—and, you know, this is one of the first times I've been in touch with some of my resources. I got up this morning almost liking myself!

Group Member F: It sounds like a very pleasant surprise. There's a lot more there than you thought!

In each case here, the person responding assimilates what the other person has said and communicates his understanding of it, but he doesn't go beyond the direct message.

When joined with respect and genuineness (which will be discussed in the next chapter), primary-level accurate empathy helps dramatically to establish rapport. People who take pains to understand one another grow to trust one another. Accurate empathy makes it easier for group members to reveal and explore themselves. Let's return for a moment to the example of Group Members E and F above:

Group Member E: I did that exercise in interpersonal strengths and weaknesses last night—and, you know, this is one of the first times I've been in touch with some of my resources. I got up this morning almost liking myself!

Group Member F: It sounds like a very pleasant surprise. There's a lot more there than you thought!

Group Member E: I'm surprised that I'm even talking as directly as I am. A couple of you whom I see as high-powered—you, Gert, and you, Jeff—I've been too uncomfortable to talk to you. I guess I've been sitting here feeling inferior, but that exercise gave me a big lift.

Member E does what someone who is understood usually does—he moves forward. In this case, he opens up the possibility of moving forward in two relationships that were going nowhere. When you "hit the mark" in your understanding of another person, you help free him or her internally. Mutual accurate empathic understanding helps free and mobilize the whole group.

Advanced Accurate Empathy (AE II)

Perhaps a better understanding of primary-level accurate empathy can be gained by comparing it with advanced accurate empathy (which will be treated more extensively under the skills of challenging). AE II gets at not only what the other person states and expresses but also what he *implies* or leaves unstated or doesn't clearly express. For instance, from time to time Peter mentions this and that dissatisfaction with his interpersonal style. He also says that others have difficulty liking him, that they find it hard to accept him. He expresses dissatisfaction with the group experience because he "can't get into it." Finally, one of the group members, after listening to and interacting with (or trying to interact with) Peter, says:

> Peter, I've heard you talk about the concrete ways in which you are dissatisfied with your interpersonal style. I've heard you say how difficult you think it is for others to accept you. I'm beginning to think that it might be very difficult for you to accept yourself. I haven't been able to get the feeling that *you* really like Peter.

This is not primary-level accurate empathy, because Peter has never said that he dislikes himself and that he finds it difficult to accept himself. However, a fellow group member has inferred this from what he has said and done in the group. As with AE I, the ultimate criterion for the appropriateness of AE II is the response it elicits from the other person.

> *Peter* (haltingly, on the verge of tears): I don't think I do like myself very much. (Pause) I can't ever remember hearing anything positive from my mother or dad. Demands, yes. Encouragement or affection, no. I feel lousy about myself.

Peter moves ahead. The issue of his not accepting himself, of his having *learned* not to accept himself, is now out in the open where it can be dealt with more creatively.

A caution: if this kind of advanced accurate empathy is used too early in a relationship or in group interaction, it can be too frightening. AE II assumes that a relatively solid relationship has already been formed. Premature AE II smacks of "playing the psychologist," since it does not arise naturally out of the relationship. It is inefficient for group members to confuse, scare, or anger one another by premature responses of advanced accurate empathy. When group members begin to explore themselves rather freely, this is a sign that it is now possible to move into deeper explorations and to challenge. Nobody in the group wants to feel that he or she is being "psyched out."

EXERCISES IN THE COMMUNICATION OF PRIMARY-LEVEL ACCURATE EMPATHY (AE I)

The following exercises go beyond discrimination (perception) to communication. Although it is true that the communication of accurate empathy must ultimately be verbal, these written exercises can act as a bridge between reading about the theory of accurate empathy and seeing a few examples and the practice of accurate empathy in face-to-face situations. The methodology followed here is (1) get the theory, (2) assimilate the examples, (3) practice at your leisure through written exercises, (4) practice orally in face-to-face situations, and finally (5) use accurate empathy in open group experience. In all of the following exercises, try to imagine yourself actually listening to the speaker.

Exercise 27: The Communication of Understanding of Feelings (One Feeling)

These are the kinds of statements you might hear in your group. Picture yourself listening to the speaker. This exercise should give you some experience of responding directly to the feelings of another.

Directions

Read the statement, pause for a moment, and then write down the description of the speaker's feelings that comes to mind immediately. Then reread the statements and check yourself for accuracy. The second time, see if you can come up with a better response to each statement. Feel free to use not only individual words but phrases as well (as suggested in the "expression of feeling" section in Chapter 2).

Note that in the next few exercises you will use the somewhat artificial formula "You feel (*word or words indicating feelings*) because (*words indicating the content, experiences, and/or behaviors underlying feelings*)." This formula will get you used to identifying both feeling and content. Later on you will be asked to do it in your own way, using your own language.

1. This is a hell of a mess! Everybody here's ready to talk, but nobody is ready to listen. Are we all so self-centered that we can't take time to listen to one another?

   a. Your immediate response: "You feel ______________________________."

   b. Your response on reflection: "You feel ______________________________."

2. You and I have been fighting each other for weeks—not listening to each other, pushing our own agendas, being competitive. I think today we did what we feared the most. We talked to each other. And you know, it's been very good talking *to* you, or *with* you, rather than *at* you.

   a. Your immediate response: "You feel ______________________________."

   b. Your response on reflection: "You feel ______________________________."

3. John, when you talk that directly and caringly to me, I almost want to get up and dance. You're a breath of fresh air! I thought you'd be the last one to challenge me. I think others have told me in a variety of indirect ways that I'm self-centered, but you're the first one who has told it to me in a way that reaches me.

a. Your immediate response: "You feel __________________________."

b. Your response on reflection: "You feel __________________________."

4. A couple of months ago, I couldn't even have imagined myself saying what I'm about to say—especially to a *group* of people. That certainly says something to me about this group. I feel a need to tell you that I'm homosexual. I don't mean that I'm offering myself as a problem, or that I want "help" here. But knowing that about me, I think, may help you understand some of my interactions here. Most important, I'm very uneasy about my sexuality. It bothers me and makes me uncertain about myself, and it's that uncertain guy whom you see here. I think I can say this now because I trust you to understand me and not to "deal with" me.

   a. Your immediate response: "You feel __________________________."

   b. Your response on reflection: "You feel __________________________."

5. I look at myself as so average. My relationships outside this group are very ordinary. My relationships inside the group are very ordinary. I'm not a colorful person. Neither am I particularly a problematic person people would rather avoid. I'm just this garden-variety me.

   a. Your immediate response: "You feel __________________________."

   b. Your response on reflection: "You feel __________________________."

6. Mike, I'm not sure, but my guess is that what I have to say will be a bit embarrassing for you. I see that I could become dependent on you, and yet I can't, because you don't allow that. But you don't push me away, either. And so, when we interact, I feel whole. Tactful demands are made on me to be mature, to stand on my own feet. This is a new interpersonal world for me, and you contribute a lot to it.

   a. Your immediate response: "You feel __________________________."

   b. Your response after reflection: "You feel __________________________."

7. Look, Carl, I've already admitted that I have a hard time getting in touch with my feelings. You keep asking me how I feel. I hate that question! It doesn't help me, and from you it sounds like a cliché. When I tell you I don't know, you get ticked off. What do you want from me?

   a. Your immediate response: "You feel __________________________."

   b. Your response upon reflection: "You feel __________________________."

8. My bet is that a lot of you see me as holding back. I'm no martyr, but I have been hurt or stung in relationships in the past. Maybe you people haven't, and you can jump right in. No one here has called me slow or chicken, but—maybe it's my paranoia—the atmosphere almost tells me that you're looking at me and asking under your breath "Where is he?" I'm slow. I admit that. I'm moving at my own pace.

a. Your immediate response: "You feel ______________________."

b. Your response upon reflection: "You feel ______________________."

9. I'm not sure that everyone here is highly committed to this group, but I'd like to say that I am. I think that up to now I've been apologizing for that—at least to myself. I don't want to apologize. I'm finding out a great deal about myself. Some of it's painful, but it's worth it. I've kept some feelings I have about myself to myself up until now, but at this point I think I'm ready to move. I can't make decisions for anyone else here, but I'm going to get my money's worth.

   a. Your immediate response: "You feel ______________________."

   b. Your response upon reflection: "You feel ______________________."

10. Sometimes, Cathy, I feel like giving you a big hug! You're so warm and understanding when you want to be. (Pause) The week before last we hit it off together, both in the group and when we stopped for a bite afterwards. But I don't know where you were last week. I made a couple of attempts to reach you in the group, and—nothing. It's been the same this week. You're not hostile, but I just don't know where you are.

    a. Your immediate response: "You feel ______________________."

    b. Your response upon reflection: "You feel ______________________."

Share your responses with a partner, a subgroup, or the entire group. Check them against the suggested responses given in the Appendix. Do your immediate responses hit the mark, or do you still need time to reflect in order to be accurate? Are your responses stilted, or are they natural? Are they you?

Exercise 28: The Communication of the Understanding of Content

This exercise is the next step. You are asked not just to identify but also to communicate understanding of the experiences and/or behaviors that underlie the speaker's feelings.

In order to enable you to focus on just the *content*, the stems below will provide the "feeling" words or phrases. You supply merely the "because" part of the response. The stimulus phrases are the same as those for Exercise 27. You are still responding to those ten statements.

1. You feel angry because ______________________.
2. You feel at peace because ______________________.
3. You feel very pleasantly surprised because ______________________.
4. You feel safe enough to risk yourself because ______________________.

5. You have a case of the "blahs" because ____________________ .
6. You feel like a new person because ____________________ .
7. You feel resentful because ____________________ .
8. You feel on the spot because ____________________ .
9. You feel full of energy because ____________________ .
10. You feel you've been left hanging because ____________________ .

Share your statements with your fellow group members and see how you might have improved your own. Are you able to get at the core of what the other person is saying without being too wordy? Check your responses with the suggested responses in the Appendix.

Exercise 29: The Communication of the Understanding of Feelings (More Than One Distinct Feeling)

When people speak to one another, they don't limit themselves to the expression of just one emotion. Often, conflicting or contrasting emotions are expressed, even in a relatively short statement. For example:

> I love him a lot, but sometimes he really drives me up a wall!

The purpose of this exercise is to help you communicate primary-level accurate empathy—first with respect to feelings—to someone who expresses two different emotional states.

Read the following statements. Imagine that the person is speaking directly to you. In this exercise, limit yourself to responding to the two distinct emotions you see being expressed (in the next exercise you will be asked to deal with content). For the present, use the formula "You feel both . . . and . . . ."

1. George, I keep telling myself not to move too quickly with you. You are so, so quiet, and when you do talk you usually start with a statement about how nervous you are. It's obvious to me that right now you're pretty fidgety and probably wish that I hadn't said anything to you. It's like a checkmate: if I move I lose, and if I don't move nothing will happen between us, and I'll lose.

   "You feel both ________________ and ________________ ."

2. Elaine, in the two weeks we've been together here my response to you has been very positive. It's a little hard to say this to you. My tendency is to get to know the men first in a social situation, since I feel more comfortable and accepted initially by men than by other women. I like you, though, and I

want to trust that feeling instead of waiting to see if you'll somehow "prove yourself."

"You feel both ____________________ and ____________________."

3. Bill, I've found you to be easygoing and rather playful. I've warmed up to you quickly and enjoyed your lightheartedness. Lately, though, I've had the impression that you use humor as a defense whenever things get intense here, and I'm a little put off by your taking things so lightly.

   "You feel both ____________________ and ____________________."

4. I'm having a hard time getting involved here tonight. So far we've only gotten into areas that have been painful, and I'm just not up to it right now. It seems as if all of my friends are coming apart at the seams, and when that happens here in the group too—(Pause) well, I'd like to be understanding and accepting, but I'd rather run like hell from everyone.

   "You feel both ____________________ and ____________________."

5. Jan, I know I come across well here in the group. In a way I'm pleased with myself. I like hearing that from you. Yet in a very strong way I don't feel pleased with myself. To be really honest, I'm just not like this anywhere else. I feel safe in this group, so it's not hard for me to be honest and direct. Outside of here I'm like a turtle in a shell. I don't take the first step to deepen a relationship, even though I ache to do so at times.

   "You feel both ____________________ and ____________________."

6. John, you're usually warm and accepting with me, but I'm still unsure of where I stand with you. I guess I want you to be very effusive in your response to me, and that's not how you are. Maybe what I'm saying is that I need a lot of attention. I know that whenever I say something I expect it to be picked up as meaningful and insightful. I'm wondering now if I've been putting too many demands on you.

   "You feel both ____________________ and ____________________."

7. You know, this is my first session as a trainer. I couldn't wait to begin this lab. I know this is the kind of work I want to do. Yet it's been downhill ever since our group began. Cathy, I could take it when the others got angry and blamed me for Sally's hurt feelings, but I didn't expect you, my co-trainer to turn on me too.

   "You feel both ____________________ and ____________________."

8. Kevin, it's difficult for me to believe that you're for real. I mean, well, you're the oldest one here and unlike anyone your age I've ever met. You seem so alive, so enthusiastic, and—best of all—you work at understanding without

making judgments. I'd like to believe you are genuine, but if that's true I'll have to change my act of staying away from older people.

"You feel both __________________ and __________________."

9. Dale, I'm really drawn to you, basically because you dare to be so vulnerable. Sometimes it's hard for me to be with you when you cry, though; the intensity of your crying frightens me and I hardly know how to respond. I want to encourage you to express yourself as freely as you can, because it helps me to take more risks, but I also want you to know that sometimes you leave me speechless.

   "You feel both __________________ and __________________."

10. Maureen, what you're saying rings true with me. I like my job and work very hard at whatever tasks I have to do. But I can also get so involved in helping people that I have little time left for being with my friends. Sometimes I wonder if I care more about other people than I do about myself.

   "You feel both __________________ and __________________."

Share your responses with your fellow group members. Check the suggested responses given in the Appendix. Did you identify the major emotions? How could you improve your responses? Try to improve on the suggested responses in the Appendix.

Exercise 30: The Communication of the Understanding of Content (More Than One Distinct Feeling)

This exercise concludes the previous exercise and asks you to "put it all together"—that is, to hook up distinct feelings with content. Distinct experiences and/or behaviors give rise to the distinct feelings.

Once you have shared your responses to Exercise 29, you have already identified the distinct emotions expressed by the speakers in the statements. Now that you have correctly identified the emotions, tie each emotion up with content, as in the example below. Continue to use the formula "You feel . . . because . . . and/but you also feel . . . because . . . ." The first one is an example.

1. You feel cautious with George because you want to respect his pace, but you also feel on edge because you're afraid that nothing is going to happen in your relationship.

Now continue to respond to the rest of the statements.

Check your responses with those of your fellow group members and with the suggested responses given in the Appendix. How accurate have you been? How "lean" are your responses? Are they too long?

Exercise 31: The Full Communication of AE I, Both Feeling and Content

This exercise is designed to achieve two goals: First, you are asked to "put it all together" and respond completely with accurate empathy—that is, to respond to both feelings and content. Second, you are asked to begin to use your own language—your own verbal style—instead of the formula "You feel . . . because . . . ." By now this formula should have begun to outlive its usefulness. Genuineness demands that you respond to others naturally, using your own style.

Directions

Read the following statements. Try to imagine that the person is speaking directly to you. You have two tasks:
a. Respond with accurate empathy (primary-level), using the formula "You feel . . . because . . . ."
b. Next, write a response that includes understanding of both feelings and content but is cast in your own language and style. Make this second response as natural as possible. The first is an example.

Group Member A: I had a hard time coming back here today. I felt that I shared myself pretty extensively last week, even to the point of letting myself get angry. This morning I was wondering what kind of excuse I could make up for not being here.

a. You're feeling awkward about being in the group tonight because—given last week—you aren't sure how I, or the others, will receive you.
b. It's not easy being here tonight. You've been asking yourself how you're going to be received. In fact, you're so uneasy that you almost didn't come.

1. John, why do you have to compare me to Jane and Sue? I do that so much myself—always trying to measure up to someone else's standards. It's something I'm really trying to break myself of. And then you come along and compare me, too.

   a. ______________________________

   b. ______________________________

2. Gary, you seem to have everything so together. You're good at all of these skills, and you even seem strong when you're talking about your vulnerabilities. Or, at least, when you're talking about some weak point I allow myself to hear only how you are on top of it, how you have it under control. And then I just take another long look at my own inadequacies.

   a. ______________________________

   b. ______________________________

3. I've known for a while that I haven't been too comfortable with sharing myself here. Cliff, when you began talking this evening about your misgivings about going into a psych internship, it hit me. I'm the only one here who isn't headed for some kind of career in a mental-health profession—and you people keep making career references. Careerwise, I'm different from all of you, and I'm beginning to wonder if it isn't more than that.

   a. ______________________________

   b. ______________________________

4. This group experience has really put new life into me. I've needed acceptance and encouragement, and I've received a lot of it. It's made me want to experiment with new behaviors—like talking about my feeling overcontrolled by others. This is really different for me.

   a. ______________________________

   b. ______________________________

5. Greg, I've been thinking about what you said about my being cynical here. It does set me apart and make me almost invulnerable. I've always prided myself on the fact that no one can get to me. It gives me freedom. I can talk to anyone and not fear that I'll get swallowed up or trampled on. I've seen this as a good setup. It never dawned on me that there might be a negative side to this coin.

   a. ______________________________

   b. ______________________________

6. You're asking me for more than I want to give. I know that groups are meant to bring people together, but there's a point beyond which I don't want to go. Demands of friendship, I think, are beyond that point.

   a. ______________________________

   b. ______________________________

7. Jane, earlier this evening I began asking you to take a look at the way you sidetrack issues that put you in center stage—and then I backed off. I couldn't read your behavior—whether it said "back off" or "proceed with caution" or what. I was getting the feeling that maybe I wasn't the best one to be making demands on you.

   a. ______________________________

   b. ______________________________

8. When you look at me with those big brown eyes, I—well—I get immobilized. A lot that you do makes me want to run, especially when you begin ques-

tioning me about intimacy. But there's something about you that makes me want to hold my ground and talk to you.

a. ______________________________

b. ______________________________

9. I'm having a hard time fighting my envy of you. Besides being attractive, you're so damned articulate, and you always seem to be in control. You don't fall off your pedestal, and no one has been able to knock you off it.

   a. ______________________________

   b. ______________________________

10. I've learned a lot of things about my sexuality in this group. I've learned that people can feel sexually attracted toward me without wanting to go any further, without pushing me at all, without making sexual demands on me. I've learned that I can begin exploring my own sexual feelings, about both men and women, without feeling like a jerk. Just because I talk about my sexual feelings, people don't immediately start thinking I'm on the prowl. Better yet, I don't see *myself* on the prowl.

    a. ______________________________

    b. ______________________________

Share your responses with others. First check them for accuracy. Next check to see how natural your "b" responses are, and ask others if these responses sound like you. Finally, see if you or others have better responses than those suggested in the Appendix.

Exercise 32: The Full Communication of AE I: Contrasting Emotions

This exercise is an expansion of the previous exercise. Therefore, it also has two goals. You are asked to "put it all together" again and respond with full primary-level accurate empathy. This time, however, the speaker will express two distinct or contrasting emotions. We don't lead simple emotional lives. We very often feel two sets of emotions in our transactions with others (for example, approach and avoidance). Accuracy demands that we be able to identify and respond to both, for responding to only one distorts the picture. Second, you are again asked to cast your response into your own language and verbal style.

Read the following statements. Try to imagine that the person is speaking directly to you. Then:

a. Respond with AE I, using the formula "You feel . . . because . . . ," keeping in mind that the speaker will express more than one emotional state.

b. Second, write a response that expresses your understanding in your own language and verbal style.
The first is an example.

Group Member A: I've never experienced anything quite like this before. I can speak my mind in this group. I can be utterly myself, and I even see myself in a kind of mirror through the feedback I get. I get encouraged to be more assertive, because that's what I want and need; but people here aren't afraid to tell me that when I become more assertive I also become more controlling. I wonder why, then, I still act a bit defensive here. Almost in spite of myself, everything I do here still says "Be careful of me."

a. You feel a great deal of satisfaction here because you can entrust yourself more fully than ever to us, and yet you feel uneasy because the trust isn't complete, and you find yourself still instinctively on guard.

b. You trust people here; you trust the direct way they deal with you, and you like that very much. But something is still making you cautious, and this need to be cautious seems to be making you uneasy.

1. I always thought that doing exercises in groups would be very phony, but I certainly can't say that about the exercise we just did. I'm still not sure that the physical touch part is really "me," but maybe it would do me good to be freer in the ways I express myself. If exercises can help me be more myself or what I want to be, well, maybe they're all right—at least some of them.

   a. ______________________________

   b. ______________________________

2. I don't know what to do with you! You look so sincere, and I believe you're sincere. I think you actually have my interests at heart. You talk to me here. You make me look at myself—my fears of getting close to others, my use of boredom as a defense. But the way you do it! You keep after me. You make the same point over and over again. Sometimes I want to run out of here screaming!

   a. ______________________________

   b. ______________________________

3. We've been in this group meeting tonight for over two hours, and —let's face it—nothing has happened. We've been tiptoeing around one another, and it gets so bland I could throw up or go to sleep. Here, let me own my own part. I see all this nothing going on, and I even watch myself letting it go on and do nothing.

   a. ______________________________

   b. ______________________________

4. Well, we have only one more group meeting. "Termination" I guess they call it, and all the stuff that goes with it. Last night I was trying to think about the good things that have happened. There's been a lot. It's not that I could count up a lot of bad points; it's that I'm not sure I'd like to live as intensively as I almost *have* to live when I'm here.

   a. ______________________________

   b. ______________________________

5. You know, Bill, sometimes I say things to you and I can see that they have an impact on you—like the time I talked to you about being cynical with the trainer. I like seeing that I affect you—maybe because you affect me so much. I might even be too easily affected by you, and the balance seems more equal when you are affected by what I say, too.

   a. ______________________________

   b. ______________________________

6. There's a wall between you and me. I'm not sure I want to break it down, but I feel I should. After all, our contract is to build relationships with one another. I feel when I come in here that I have an awful lot of work to do with you.

   a. ______________________________

   b. ______________________________

7. I listen to you. I attempt to understand you. I want you to talk to me. What do I have to do to show you that I care about you? I'm not trying to own you, but I'd like some kind of response from you. Maybe it's me. Maybe I go about things in the wrong way. God knows I've done that in the past.

   a. ______________________________

   b. ______________________________

8. I've never been in a group like this before, so I don't know quite what to expect. I've heard some horror stories, but I tend to discount them. A number of my friends have taken this course, and not only do they speak well of it, but I think they did pick up some very specific skills. I'm looking forward to learning some myself. It's just that it's so new.

   a. ______________________________

   b. ______________________________

9. One thing that clogs up the interaction in this group is the pairing that goes on. Jane, you tend to agree with everything Fred says—although that's an exaggeration. Bill, you and Tom pair up. And Betty and I tend to form a

team. We've all got our defenders, so what place is there for risk? I'm wondering whether it's too late—we have only three weeks left—to do much about it. Shall we just settle for what we have?

a. ____

b. ____

10. Hey, we got rid of one of our big bugaboos this evening! We stopped spending endless time talking about last week's interaction. Boy, any time we get a little action going here, we kill it with post-mortem analysis. Bob, you and I have actually been talking very directly to each other. I had no idea that you'd been resenting my control, my constant whipping the group into shape. I mean I had no idea.

a. ____

b. ____

Exercise 33: The Practice of Primary-Level Accurate Empathy in Everyday Life

If responding with accurate empathy is to become part of your natural communication style, you will have to practice it outside the formal training sessions. If accurate empathy is relegated exclusively to officially designated helping sessions, it may never prove genuine. Actually, practicing accurate empathy is a relatively simple process.

1. Begin to observe conversations between people from the viewpoint of the communication of accurate empathic understanding. Does a person generally take the time to communicate this kind of understanding to another person? Try to discover whether, in everyday life, the communication of accurate empathic understanding (primary-level) is frequent or rare. As you listen to conversations, keep a behavioral count of these interactions (without changing your own interpersonal style or interfering with the conversations of others).
2. Try to observe how often you use the communication of accurate empathic understanding as part of your communication style. In the beginning, don't try to increase the number of times you use accurate empathy in day-to-day conversations. Merely observe your usual behavior.
3. Increase the number of times you use accurate empathy in day-to-day conversations. Again, without being phony or overly preoccupied with the project, try to keep some kind of behavioral count. Use accurate empathy more frequently, but do so genuinely. You will soon discover that there are a great number of opportunities for using accurate empathy genuinely.
4. Observe the impact your use of primary-level accurate empathy has upon

others. Don't use others for your own experimentation, but, once you increase your use of genuine accurate empathy, try to observe what it does for the communication process.

SOME COMMON PROBLEMS IN COMMUNICATING PRIMARY-LEVEL ACCURATE EMPATHY

The communication of primary-level accurate empathy looks simple on paper, but this simplicity is deceptive (as you will find out as you practice this skill). When AE I is called for in communication, you will find tendencies in yourself to give other, less facilitating, responses. Let's take a look at some of these.

The cliché. If you respond to a person's self-disclosure with a cliché, this response is usually worse than none at all, for there is something demeaning about it. Clichés put distance between communicators. For example, Group Member A says, somewhat nervously:

> We haven't talked about sexual feelings here at all. If I am honest with myself, I have to say that I'm not sure where I stand sexually at all. I've thought about it a lot during the group meetings, but I haven't said anything. And I've been very cautious with the women here. I think that my guard has been up, and I bet some of you have noticed it. I feel dumb because at my age I'm not supposed to be sexually immature, but I think I am.

Consider the following responses:

> *Group Member B:* Many people struggle with sexual identity throughout their lives, John.

This may sound silly here, but don't laugh. Responses like this are made in groups. Or:

> *Group Member C:* John, I think I understand.

This commonly heard cliché is often simply a lie. The person responding does *not* understand, but he feels that he should say something. Clichés don't help the speaker move forward at all.

The question. We will explore the use (or nonuse) of questions further later on, but questions are generally very poor substitutes for accurate empathy. If, in response to John's statement concerning his sexual confusion, a group member asks:

> In what ways do you think you are sexually immature, John?

he ignores the fact that John just risked himself. Reasonable risk-taking adds a great deal of life to groups, but risk-taking is possible only in a climate of support. And the communication of accurate empathic understanding is at the core of support. Questions are useful if they help another person probe an area; an "open" question is one that does this. Once a person explores himself, he should be understood.

Premature advanced accurate empathy. Perhaps I should say "premature and inept" advanced accurate empathy. For instance (again, in response to John's statement about his sexual confusion):

> *Group Member D:* This sexual thing is probably just a symptom, John. I'll bet something else is really bothering you.

This misplaced (and poor) attempt at AE II turns Member D into the group "psychologist." He might even compound his mistake by saying:

> Don't you see? Sexual immaturity is just another way you try to hang onto your childhood and its safety.

Adding psychodynamic interpretations of highly questionable validity merely compounds the crime and creates greater distance between the two communicators. Primary-level accurate empathy is an excellent way of becoming meaningfully present to another.

> *Group Member E:* It sounds as if sexuality has been a rather disturbing issue for you. It might even help if we all would take a look at how sexuality affects us here.

Member E's response is the most facilitative of all, because it shows understanding for John's feelings and an initial willingness to see whether his concerns might be dealt with in the group.

Inaccuracy. Accurate empathy should be precisely that—accurate. If your understanding of another is inaccurate, he will probably let you know in one of a variety of ways: he may stop dead, fumble around, go off on a new tangent, tell you "that's not exactly what I meant," or use some other means to show that your response has blocked him in some way. An example:

> *Group Member A:* Maureen, I don't mean to imply that you're always butting into my life here. There are times when there is some mutuality between us, when we cooperate, when we explore what's going on between the two of us—or what isn't going on.
>
> *Maureen:* You feel frustrated by the controlling way I interact with you. I intrude on you rather than interact with you.

> *Group Member A:* Well, that may be. But I'm not sure that that's what I just said.

Maureen isn't accurate. She is pursuing her own agenda rather than listening and responding to the other. Perhaps she is trying to play the game of "catch-up." The response she gives here might have been an excellent one two interchanges previously, but it doesn't fit here. "Catch-up" and good will are no substitutes for here-and-now accuracy. Maureen could have said:

> There *are* times when there is give-and-take between you and me. And there's something satisfying about this.

Here she responds to what was said, and her AE I is a sign of mutuality and cooperation instead of control.

Feigning understanding. Sometimes it will be difficult for you to understand what the other person is trying to communicate, even though you attend to him quite fully. This will happen especially often if AE I is not already a part of your everyday communication skills. If the other person is confused, distracted, or in a highly emotional state, the clarity of what he is saying may be affected. Or you yourself may become distracted, lost in your own thoughts, or confused, and fail to follow what the speaker is saying. At such times, don't feign understanding. This is phony. Genuineness demands that you admit that you are lost and that you work to get back on the track again. "I think I've lost you. Could we go over that once more?" If you're confused, admit your confusion. "I'm sorry. I don't think I got what you just said straight. Could you go through it once more?" Such statements are signs of your respect, of the fact that you think it is important to stay with the other. Admitting that you are lost is infinitely preferable to such clichés as "uh-huh," "ummmm," and "I understand."

If you are not quite sure that you understand what the other person is trying to express, be tentative in your responses and give the other "room to move." He will then feel free to correct you or give a clearer picture of what he means.

> *Group Member A:* You seem to be saying that I find it difficult to trust you because your emotions change so rapidly, even during the same group meeting. Is it something like that?
>
> *Group Member B:* That's partly it. I think that I don't even trust *myself* when my mood changes so rapidly. I don't expect to be trusted.

A tentative response has given this person (B) the opportunity to clarify what he means, both for the person to whom he is talking and for himself. A does not hem B in by what he says; rather, he engages in "perception checking" (Brammer, 1973, p. 86), which keeps the communication open.

Perception checking means that you ask the person with whom you are talking for verification of the accuracy of your response or series of responses. Asking for confirmation from the other is a way of indicating that it is important for you to understand him. As Brammer (1973) notes, in ordinary social conversation "we are conditioned to chatter on socially, even deliberately confusing the meaning with innuendo, humor, and metaphor. We rarely check with one another about what we are really trying to say" (p. 86). There are a number of ways of going about checking the accuracy of your understanding. Here are some examples. The italicized words are part of the perception-checking process.

- It *sounds like* you're angry with both the group and yourself, *but I'm not sure.*
- It *seems to me* that you'd like a little time to think about what has just happened. *Am I reading you correctly?*
- You *seem to be saying* to Joan that you like her but that she also does things that annoy you. *Is that the way it appears to you?*
- You *seem to be* very relieved that you and Bill are talking to each other freely again. *Or is that putting it too strongly?*
- I *have the feeling* that you're angry with me and are keeping it back. *But maybe I'm reading you wrong.*

Perception checking helps keep the dialogue flowing freely. It rids statements of accurate empathy of overtones of judgmentalness and thus is a sign of respect.

Parroting. Parroting is a mechanical restatement of what another person says.

> *Keith:* I just dumped my anger on you, but I don't think that was fair. I'm actually angry at myself for running away from confrontations that are quite legitimate. Attacking is one way I run.
>
> *Bill:* You don't think it was fair to dump your anger on me. You're really angry with yourself for running away from legitimate confrontations. Attacking me was a form of running.

Accurate empathy means that I make what the other says "my own" and then respond. The other's experiences, behaviors, and feelings get "inside" me in some way, and my response is a giving of myself. Parroting can be done by a tape recorder (and done most efficiently); there is usually little mutuality or human contact in parroting. The good communicator is always looking for the core of what is being expressed by the other person. The good response gets at the essence of what the other person is saying and is not merely paraphrase or repetition.

Jumping in too quickly or letting the other ramble. Good communication demands that you give other people an opportunity to say what they want without letting them just ramble. Overeagerness is awkward and disconcerting to the person speaking. Giving the other a chance to speak gives you a chance to ask yourself: What feelings are being expressed? What point is this person trying to make?

At the same time, spontaneity is valuable. You can speak up any time you have something you really want to say to another, even if you have to "interrupt" him. I put "interrupt" in quotation marks because, in my experience, group members are overly polite with one another. They allow one another to ramble, using the pretext that they don't want to interrupt. A spontaneous interruption can be a very helpful interaction. The high-level communicator interrupts in a non-self-conscious way. He is so "with" the person speaking that he doesn't have to ask himself whether he should speak or remain silent. Not interrupting (in the sense in which it is being explained here) is often more closely associated with fear or disinterest than with the welfare of the speaker.

Discrepancy in language, tone, and manner. You probably communicate more effectively if your language is in tune with the language of the person to whom you are speaking—or if the two of you can at least find a common language ground. For instance, if in your group your language is stilted and professional whereas another's is casual and laced with the argot of the day, language itself will keep you away from each other.

> *Group Member A:* Man, I'm spaced out right now. Not sure you cats even know where I'm coming from. I'm about to blow my cool. This is a psychological rip-off.
>
> *Group Member B:* You feel frustrated and angry because our perceptions of you are not really empathic.

These two people are worlds apart in terms of language. You can be yourself and still accommodate in some degree to the language of the person you are trying to understand. Let's have B try again:

> You feel that nobody here's picking you up, and that's a lousy feeling.

Here he does not mimic or imitate a language that is simply not his own, but he gets away from the formalism of his previous statement. In terms of language he is closer, more empathic. In practice sessions, such formulas as "you feel . . . because . . ." are useful, for they remind you to focus on both the feelings and the underlying experiences and behaviors of the other person. After awhile, however, they lose their usefulness because they often don't represent you and the way you usually talk. As you become more skilled and more comfortable with the communication of accurate empathy, you will find yourself more capable of using language that is both your own and in tune with the person with whom you are communicating.

If one of the members of your group speaks animatedly, telling of her elation in learning skills and establishing some solid relationships with some of the other members of the group, and you respond in a flat, dull, lifeless voice indicating how glad you are to share her joy, your response loses a great deal of its empathy—even though you may well be accurate in identifying her feelings and the experiences and behaviors underlying those feelings. She will hear the nonverbal and paralinguistic messages and find you rather nonempathic. This does not mean that you should simply mimic her enthusiasm (or anyone's emotional expression), but there should be some proportion between your tone and manner of responding and those of the speaker. Ultimately, empathy means being with the other in a non-self-conscious way. The "being with" is primary, and the process of communication should conform to this primary goal.

Longwindedness. There should be a great deal of *dialogue* within the group. Longwinded monologues followed by longwinded analytic responses are deadly. Responses can often be improved if they are made lean and to the point. It usually takes more words to say nothing than to say something. An example: two members of a group have spent a great deal of energy avoiding each other. When they finally begin talking, they feel awkward; they don't know what to say. Finally:

> *Group Member A:* I like you, but with the qualification "as far as I know you." I don't know you too well, but I like what I see.
>
> *Group Member B:* I like you without qualification.

B's remark is short, but it has great impact.

In trying to be accurate, the beginner may tend toward longwindedness, sometimes speaking longer than the person to whom he is responding. This often happens when he responds too quickly, without sufficient reflection. He begins to speak, realizes that his first few sentences have been wide of the mark, and then keeps on talking in hopes of eventually hitting it.

> *Group Member C:* I've never been very spontaneous in social situations, and this one is no different. I'm shy—in the sense that I stand off to the side, at least figuratively, and wait to see how I can get into the conversation. As a result, more often than not the conversation passes me by. By the time I'm ready to say something, the conversation has gone someplace else, and my contribution is no longer relevant. I've been inside myself and not in touch with what's going on.
>
> *Group Member D* (jumping in right away): You're really shy, and that cuts down on your spontaneity. It shows up in groups like this one. You're listening, all right; you know what people are saying. But then you begin to ask yourself

"What should I say? I shouldn't just sit here dumb like this." But by the time you think of what to say, it's just too late. No, it's worse than that. Now you've lost the thread of the conversation and it's twice as hard to get back in. Your shyness backfires on you in more than one way. And you sit here even more alienated.

This response may well be accurate, but it is too much. It places emphasis on D's attempts to understand rather than on what is bothering the speaker. C is smothered by all that D has to say, and therefore finds it difficult to move forward. If you keep asking yourself "What is the core of what this person is saying?" until it becomes second nature, you can make your responses short, lean, concrete, *and* accurate.

RESPONDING TO FEELINGS OR TO CONTENT

In the examples given so far and in the exercises you have been asked to do, response to *both* feelings and content (the experiences and behaviors underlying feelings) has been emphasized. Ordinarily, this is the best kind of AE I response. However, at any given time, one or the other may be more important. For example:

Group Member D: What's going on here? We're all good at these skills, but we're still on the same damn plateau. I build relationships with you people, but they seem lifeless. People respond well to me, but there's still a lot of distance between us. I don't feel any movement in the group, and I know it's at least partly my fault.

Group Member E: You're really fed up.

Group Member D: You're damn right I'm fed up. And I'm hoping that I'm not the only one who's fed up. I want to move, and I think a lot of you do, too.

E responds only to D's feelings, but his response is empathic enough to encourage D to move along. E sees feelings in this case as the core of what D is saying. Another case may be quite different.

Group Member F: Norm, when Jean is silent, it bothers you a lot and you tell her so pretty quickly. But when I'm silent, you give me time. You let me move along at my own pace. I think I can name other instances in which you deal with people here quite differentially.

Norm: I must seem pretty inconsistent to you.

Group Member F: Yes, you do, but I suppose what bothers me the most is that I don't know on what you base this differential treatment. I don't see you working out your relationships concretely enough to get a feeling for what you are doing.

Norm listens, and he responds nondefensively, but to content (F's experience) rather than to feelings. Here, the content seems to be the core of what is being said. F hits the mark, and Norm moves further along in revealing what he sees happening. As usual, the criterion for a good response is whether it encourages concrete dialogue. In this case it does.

If a group member shows early in the life of the group that he is threatened easily by discussion of his feelings, you may want to emphasize content with him in the beginning and proceed only gradually to dealing with feelings. This approach constitutes a kind of process of desensitization for such a group member. One relatively nonthreatening way to encourage another to deal with feelings is to indicate what you might feel in similar circumstances.

> *Group Member G:* A lot of the responses I get here seem to assume that I'm a little kid. And I'm in my mid-30's! I get encouraged to talk. I get gentle reprimands. I feel that my lack of initiative is tolerated.
>
> *You:* I think that if I were being treated that way here, I'd be pretty angry.

This response, however, could backfire if G sees it as just another instance of parental behavior; it is just such parental behavior that is bothering him.

Exercise 34: The Identification of Common Mistakes in Phase I

The following exercise deals with some of the common mistakes people make when responding to another person. These faults or mistakes consist, in effect, of poor execution of primary-level accurate empathy. Before you do the exercise itself, let's review briefly what some of these common mistakes are:

- responses that imply condescension or manipulation
- unsolicited advice-giving
- premature use of advanced-level accurate empathy
- responses that indicate rejection or disrespect
- premature confrontation
- patronizing or placating responses
- inaccurate primary-level empathy
- longwindedness
- clichés
- incomplete or inadequate responses (such as "uh-huh")
- responses that ignore what the person said
- use of closed, inappropriate, or irrelevant questions
- use of inappropriate warmth or sympathy
- judgmental remarks

- pairing or side-taking
- premature or unfounded use of immediacy
- defensive responses

This list is certainly not exhaustive. Can you think of other mistakes? Some of these errors are demonstrated in the exercise that follows. You are asked to identify them.

However important it is to understand a person from his own frame of reference, there is still a tendency *not* to do so—to do many other things instead. One function of this exercise is to make you aware of the many different ways in which it is possible to fail to communicate basic empathic understanding to others.

Directions

Below are a number of statements made by various group members, followed by a number of possible responses.

a. First, if the response is good—that is, if it is primary-level accurate empathy—give it a plus (+) sign. However, if it is an inadequate or poor response, give it a minus (—) sign.

b. Second, if for any reason you give the response a minus (—), indicate briefly *why* it is poor or inadequate (disrespect, premature confrontation, defensiveness, judgmentalness, condescension, and so on). A response may be poor for more than one reason. Make your reasons as specific as possible.

Study the example below and then move immediately to the exercise.

Example

Group Member A: I have high expectations of this group. I think we've developed a pretty good level of trust among ourselves, and I'd like to start taking greater risks. The longer I'm here the more desire I have to learn as much as possible about myself. I want you to help me do this, and I want to do the same for you.

a. (—) Hey, I wish you wouldn't speak for me. I'm not at all sure that my expectations are the same as yours. I think you're being pretty idealistic.
   Reason: defensive, judgmental, accusatory.

b. (+) Your enthusiasm is growing. There are a lot of resources here, and you'd like to take advantage of them.
   Reason: (none because it is a plus).

c. (—) Do you think we're ready to do this sort of thing?
   Reason: inappropriate, closed question; vague.

d. (—) Now, John, you've always been a good member, very eager; I appreciate your eagerness very much, but *festina lente,* as the Romans said—"make haste slowly."
   Reason: condescending, parental, advice-giving.

e. (+) Your enthusiasm's infectious, John—at least for me. I think that I,

coward that I am, am ready for a bit more risk, myself.
Reason: (none because it is a plus) ________________.

1. I didn't feel right barging in on Paul and Marie's conversation, so I waited until I thought they were finished. I keep thinking that people will get angry if I interrupt. It may be the wrong way to be, but I don't interrupt people outside the group, and it's hard for me to think that it's okay here.

   a. ( ) It seems that you're afraid of being rejected if you interrupt. And rejection really hurts you, because you don't see yourself as a worthwhile person.

   Reason: ________________.

   b. ( ) I think that's pretty unfair of you, since you don't give Paul and Marie much credit.

   Reason: ________________.

   c. ( ) Peter, you know the contract here. What you call "barging in" is merely "owning" one of the interactions. I know you're timid, but I think you should push in anyway.

   Reason: ________________.

   d. ( ) Direct "owning" of another conversation just doesn't seem right to you—so it's really hard for you to move in.

   Reason: ________________.

2. I think my skill level has improved significantly within the last two or three weeks. I'm able to express my feelings much more openly and honestly, and feeling more confident has helped me to become less defensive.

   a. ( ) You feel half-finished because you haven't been able to lick your defensiveness completely.

   Reason: ________________.

   b. ( ) Yeah. I can see that.

   Reason: ________________.

   c. ( ) How have you managed to become less defensive?

   Reason: ________________.

   d. ( ) I know your skills are improving, but I can't say that I see you as less defensive.

   Reason: ________________.

3. I think of myself as a pretty independent guy, and my independence, if I'm not mistaken, tends to rub people the wrong way. People figure there's no way that I can be affected by them. I'm not saying that this is right or wrong, but

I think it's only fair to let you know about it, so that you don't look for something I don't usually give.

a. ( ) I know exactly what you mean, Joe. I used to think that I was more "together" than anyone else I knew. Other people didn't matter; often enough they just got in my way.

Reason: ______________________________ .

b. ( ) The reason why we can't influence you, Joe, is that you won't let us. I'll bet that somewhere along the line you let someone get close, and you got stung. I know it's going to be hard, but you have to risk yourself like everybody else. I think you can trust us enough to do this.

Reason: ______________________________ .

c. ( ) It almost sounds as if you're saying that you see yourself as the kind of person who can alienate people by being detached. I'm not sure how you feel about it—except that you think it might cause some trouble here.

Reason: ______________________________ .

d. ( ) I don't see you that way at all, Joe. I really experience you as open and "with" us. I like being in the group with you.

Reason: ______________________________ .

e. ( ) Well, isn't that nice! You just want to be left alone—probably because you're afraid of being dealt with.

Reason: ______________________________ .

4. Janet, I'm having a hell of a time understanding you, or trying to understand you. It could be that I'm putting up a wall between us, because I know I've been defensive with you a couple of times. I get the idea that you avoid talking with me at times, but I'm reluctant to say that it's your problem. I guess I don't know whether it's you or me or both of us.

a. ( ) It's good to bring things like this up. I think I understand.

Reason: ______________________________ .

b. ( ) I like the way you can look at yourself and admit your own faults and your part of what goes wrong in a relationship.

Reason: ______________________________ .

c. ( ) You're kind of confused right now because you don't know how to deal with my aloofness.

Reason: ______________________________ .

d. ( ) I think it's rather unfair, putting it on me like this. I try not to run away from facing any relationship. I think my behavior in this group will attest to that.

Reason: ____ .

5. I don't believe what's happening to me! I came in here six or seven weeks ago expecting to "rap" or "relate" or something like that, and now I find my whole interpersonal style changing. The most satisfying thing is the change that's taking place between me and my husband. I never expected so much out of a course.

a. ( ) I'd like to take a look at what's happening between you and me, Ann. I feel that there is still some unresolved hostility between us that should be taken care of first before going into what is happening outside.

Reason: ____ .

b. ( ) I'm not sure that I understand. How exactly has your marriage changed because of the group? And I'm not sure what you mean by "rap" and "relate."

Reason: ____ .

c. ( ) What's happened here is really surprising. Especially that it could affect your closest relationship so positively!

Reason: ____ .

d. ( ) I feel left out. I know this group experience has been great for you and that you're really pouring yourself into it. But I don't feel that way, and I think you and I have got to go a lot further if we are going to relate in any meaningful way here.

Reason: ____ .

6. I'm not a very understanding person, and I can't change overnight. I don't see why we have to stress understanding so much. It's phony. People don't talk like that to one another. I don't see what's wrong with free-flowing conversation. It's getting so stilted in here, I wonder whether the course has any application to relationships outside. You people don't use "accurate empathy" any more than I do. Why don't we admit it and get on with relating?

a. ( ) I think you've hit the nail on the head. If our attempts at accurate empathy are so artificial, they can't do much good for us in our day-to-day relationships.

Reason: ____ .

b. ( ) Please, Mike. I wish you wouldn't push your view of things so much. I get a lot out of these practice sessions. You don't like "accurate empathy" because it's really hard for you to do well. At least don't dump your problems on the rest of us.

Reason: ____.

c. ( ) This group isn't real life, and practicing "accurate empathy" and stressing it so much doesn't seem realistic, since, if we admit it, we don't use it very much outside. So you're frustrated. You'd like to see us get down to some more realistic issues. You'd feel more comfortable with less practicing and fewer rules. It's not that you're saying that the group has nothing to offer. Just relating is okay, and you'd like to do more of it.

Reason: ____.

d. ( ) I find I'm using more and more accurate empathy, both at home and at work. I know you're frustrated, Mike, and that you see this as artificial, but I'm not sure you're giving it a fair chance—either here or outside.

Reason: ____.

7. I'd like to start by saying that it's a brand new experience for me to receive criticism and confrontation in such a way that I don't have to get defensive. I'm used to defending myself, but most of you challenge me in ways I really like. I like this group.

a. ( ) Good. (Pause) Aren't there some unresolved issues from last week?

Reason: ____.

b. ( ) Bill, you said "most of you." Who in here challenges you in non-productive ways?

Reason: ____.

c. ( ) The give-and-take in here has been pretty constructive—and that's satisfying.

Reason: ____.

d. ( ) I think I like confrontation just about as much as you do, Bill, but I haven't heard much of it from you. How about giving me some feedback?

Reason: ____.

8. This is very hard for me to say, Phil, because I admire you and like you. I'm angry with you because I had lofty expectations of you, and I feel let down. I see you as an intelligent, influential person, and I've been expecting you to help this group go places. I'm disappointed with the group as a whole, and I guess mostly with you.

a. ( ) Well, I don't think that's very fair of you. I came in here the same as everybody else.

Reason: ____.

b. ( ) You've been whining about one thing or another ever since this group began. I wish you'd get off it!

Reason: ____ .

c. ( ) This whole experience has been somewhat of a letdown for you. And maybe I've contributed more than my fair share to that letdown because you see in me resources that say I could have done better than I have.

Reason: ____ .

d. ( ) Okay, I can handle that. I appreciate the feedback. It's good for me to know how I come across to others.

Reason: ____ .

ACCURATE EMPATHY: MORE THAN A SKILL

We have been describing accurate empathy as a skill within the technology of communication, but it is more than that.

The Difficulty of Entering Another's World

The skill of accurate empathy should flow from your actually being with another person, from your experiencing his world. Sometimes people learn the technology of the skill and become rather adept at this technology, but something is missing. The "being with," I believe, depends ultimately on your ability to care about others, to move away from the self-centeredness to which we are all subject, and to experience another's experiencing. This ability, as Huxley (1963) suggests, is difficult to attain.

> We live together, we act on, and react to, one another; but always and in all circumstances we are by ourselves. The martyrs go hand in hand into the arena; they are crucified alone. Embraced, the lovers desperately try to fuse their insulated ecstasies into a single self-transcendence; in vain. By its very nature every embodied spirit is doomed to suffer and enjoy the solitude. Sensations, feelings, insights, fancies—all these are private and, except through symbols and second hand, incommunicable.
>
> Most island universes are sufficiently like one another to permit of inferential understanding or even empathy or "feeling into."...
>
> To see ourselves as others see us is a most salutary gift. Hardly less important is the capacity to see others as they see themselves . . . [Huxley, 1963, pp. 12–13].

Accurate empathy at its fullest, as a way of relating instead of just a communication skill, is an attempt to penetrate this metaphysical aloneness of the other.

The Experience of Feeling Understood

We have been emphasizing empathy from the viewpoint of the person who is trying to understand, but we should not lose sight of the effect of empathy, feeling understood. Most of us have this liberating experience too infrequently. Van Kaam (1966) did a study on the experience of feeling understood. He used a phenomenological approach, rather than the traditional empirical one, to try to determine "the necessary and sufficient constituents of the experience of really feeling understood" (p. 323). Nine constituents were finally identified as being, in combination, necessary and sufficient:

- perceiving signs of understanding from a person
- perceiving that a person co-experiences what things mean to me
- perceiving that a subject accepts me
- feelings of satisfaction
- feeling relieved initially
- feeling safe in the relationship with the person understanding
- feeling safe experiential communion with the person understanding
- feeling safe experiential communion with that which the person is perceived to represent [see Van Kaam, 1966, pp. 325–327].

Experiential communality in feeling understood provides a basis of support in the group that enables individual members to embark on the risk/experiencing-of-trust/risk cycle essential to growth.

Trustworthiness and Trust

All who write about human-relations-training groups emphasize the need for trust. Unless group members trust one another in at least some minimal way, the group learns very little. Significant learnings are almost inevitably related to high degrees of trust. Trust can be defined in a number of ways:

- It can mean *confidentiality*—that the members of the group will not reveal what goes on inside the group to people outside the group.
- It can refer to *reliability:* "Interpersonal trust is defined here as an expectancy held by an individual or a group that the word, promise, verbal, or written statement of another individual or group can be relied on" [Rotter, 1971, p. 444].
- It can refer to *consideration in the use of power,* that another will not abuse whatever power he has with respect to me. If I *entrust* myself to another, he will act with care toward me.
- It can refer to *empathy.* I expect that the other will make an effort to understand my world and achieve at least minimal success in doing so. Ultimately, then, trust is possible in the group only if there is mutual acceptance and support.

You should expect all of these forms of trust to be operative in your group. Since the total experience is governed by a contract, you should be able to expect that others will expend whatever energy is necessary to live up to the provisions of the contract.

The relationship of trust to trustworthiness is much like that of the chicken to the egg. Perhaps it is best to say that you should expect your fellow group members to be trustworthy. Trusting another is a strength, not a weakness: "The high truster does not trust out of a need to have someone else take care of him, and apparently he is not regarded (by others) as someone who is easily fooled, tricked, or is naïve" (Rotter, 1971, p. 447). The high-level communicator not only trusts but also makes himself trustworthy. Some people, such as doctors and clergymen, are seen as trustworthy because of their role; others can be seen as trustworthy because of reputation. In your training group, however, you can depend on neither of these forms of trustworthiness. What's left?

Behavior trustworthiness. Your behavior in the group becomes the most important source of your perceived trustworthiness. You can demonstrate that you are trustworthy by

- maintaining confidentiality;
- living up to the provisions of the contract;
- manifesting genuineness, sincerity, and openness;
- demonstrating respect by means of appropriate warmth, interest, availability, and cooperation with your fellow group members;
- being realistic but optimistic about the learning potential of the group experience;
- working to expand your ability to be empathic and to communicate understanding to others;
- using whatever social-influence power you have carefully, in the interests of others;
- being open to reasonable social-influence attempts on the part of your fellows; and
- avoiding behavior that might indicate the presence of such ulterior motives as voyeurism, selfishness, superficial curiosity, personal gain, or deviousness.

Kaul and Schmidt (1971) have found that a person is trusted if he respects the needs and feelings of others, offers information and opinions for the benefit of others, generates feelings of comfort and willingness to confide, and is open and honest about his motives. Others (Hackney & Nye, 1973) have found that the "underparticipating" group member is seen as untrustworthy. The other group members don't know what he is thinking and see him as judgmental in his silence. Group members also resent the underparticipating member because he is not living up to the contract, even in a minimal way, and is, therefore, unreliable.

A final caution. Primary-level accurate empathy is only one of several interpersonal skills, but it is an extremely important one. Presented by itself here, separate from the other skills and out of the context of free-flowing dialogue, it seems a bit stilted. The examples and the exercises may seem to portray a "helper-helpee" picture. The give-and-take flavor that characterizes group dialogue at its best is lost. But the point is that the communication of empathic understanding is a skill that is very underused in dialogue. If the lab were to do nothing more for you than help you weave more empathic understanding into your interactions with others, its success would be significant. The communication of accurate empathic understanding does not in itself make a nondirective counselor of you (see Rogers, 1951, 1961). It does make you more fully human in your communication. A group in which the communication of empathic understanding is a significant part of the group culture has the basic ingredient needed for building support and trust.

EXERCISES IN TRUST

Exercise 35: Physical Trust: The "Trust Walk"

This exercise will let you see what it is like to entrust your physical safety to another person; you will get a deeper feeling for the process of entrusting.

1. Choose a partner.
2. Either you (A) or your partner (B) closes his or her eyes and keeps them closed during the exercise.
3. The "blind" person entrusts himself to the other physically, allowing himself to be led around the room or down corridors.
4. The "seeing" person can have the "blind" partner do a variety of things—walk around objects, run down corridors, climb on chairs, and so on. The "seeing" partner, however, in no ways allows the "blind" partner to stumble, run into things, fall, or hurt himself in any way.
5. Allow five minutes or so for the "seeing" partner to lead the "blind" partner around.
6. Now A and B should trade roles and continue the exercise for another five minutes or so.

Discuss with your partner how it felt to entrust yourself physically to him or her. Did you move easily with your partner, or were you very hesitant? How did you handle your fears? Did anything happen that weakened your trust or the process of entrusting?

Discuss how you feel about entrusting yourself to the group in other ways—for instance, through self-disclosure. Is it easy for you to entrust yourself to others psychologically? What is easy about it? What is hard about it? What do you have to see in another person to allow yourself to trust him? How do you feel about entrusting yourself to this group of people?

Exercise 36: Physical Trust:
"Being Passed Around"

This exercise will help you see what it is like to entrust your physical safety to a group of people and get a deeper feeling for the process of entrusting yourself to a group of people.

1. Five or six members of the group form a tight circle, all members standing.
2. Another member of the group stands in the middle, encircled by fellow group members.
3. The member in the middle "free falls" toward the members encircling him or her.
4. The members in the circle don't allow the person to fall to the floor or to hurt himself in any way. The members hold the falling member up and pass him from member to member (or to two members if one member cannot assume the total burden). The members cooperate in passing the falling or limp member around the group.
5. Each person takes his or her turn being passed around.

After each member has had a turn, the members sit down and discuss the experience. How did you feel falling? Being passed from member to member? How easy or difficult was it for you to entrust yourself physically to the group? How did you feel about being touched by the group members as they passed you from person to person?

For you, is there any relationship between this kind of entrusting and entrusting yourself psychologically to this group of people? What difficulties do you have in entrusting yourself psychologically to them? What rewards do you find in entrusting yourself to them?

CHAPTER 7: FURTHER READINGS

On accurate empathy and supportive behavior:

The Counseling Psychologist, 1975, 5 (2). Half of this issue is dedicated to discussions of accurate empathy by Carl Rogers and others.

Carkhuff, R. R. *Helping and human relations.* Vol 1. New York: Holt, Rinehart and Winston, 1969.

Carkhuff, R. R. *The art of helping* (2nd ed.). Amherst, Mass.: Human Resource Development Press, 1973.

Carkhuff, R. R. *The art of helping: Trainer's guide.* Amherst, Mass.: Human Resource Development Press, 1975.

Egan, G. Supportive behavior. Chapter 8 in *Encounter: Group processes for interpersonal growth.* Monterey, Calif.: Brooks/Cole, 1970. Pp. 246-286.

Gazda, G. M. *Human relations development.* Boston: Allyn & Bacon, 1973.

Mayeroff, M. *On caring.* New York: Perennial Library (Harper & Row), 1971.

Sydnor, G. L., Akridge, R. L., & Parkhill, N. L. *Human relations training: A programmed manual.* Minden, La.: Human Resources Development Institute, 1972.

On trust:

Gibb, J. R. Defensive communication. *Journal of Communication,* 1961, *11,* 141–148.
Gibb, J. R. Climate for trust formation. In L. P. Bradford, J. R. Gibb, and K. D. Benne (Eds.), *T-group theory and laboratory method.* New York: Wiley, 1964. Pp. 279–301.
Hampden-Turner, C. M. An existential "learning theory" and the integration of T-group research. *Journal of Applied Behavioral Science,* 1966, *2,* 367–386. (There is a revised version of this model in Chapter 3 of Hampden-Turner's *Radical man.* Cambridge, Mass.: Schenkman, 1970. Pp. 31–65.)
Kaul, T. J., & Schmidt, L. Dimensions of interviewer trustworthiness. *Journal of Counseling Psychology,* 1971, *18,* 542–548.
Strong, S. R., & Schmidt, L. Trustworthiness and influence in counseling. *Journal of Counseling Psychology,* 1970, *17,* 197–204.

Chapter 8

Genuineness and Respect as Communication Skills

Although genuineness and respect can be seen as moral qualities or interpersonal attitudes, we will examine them from a different perspective in this chapter. As discussed here, both refer to *a set of behaviors* essential to a high-level communication process. Whether genuineness and respect are interpersonal *values* for you is an issue worth considering in the lab. Genuineness and respect are not innate human qualities; you have to work hard to develop them in yourself. On the supposition, however, that they *are* values for you, how are these values translated into effective communication? This chapter attempts to answer that question.

GENUINENESS: A BEHAVIORAL APPROACH

Being genuine in your transactions with others has both positive and negative implications: it means doing some things and not doing others.

Freedom from role. The good communicator does not take refuge in a variety of roles. Relating deeply to others should become part of your life-style, not a role that you put on or take off at will. Interpersonal communication at its best is role-free. How does the role-free person act in his relationships to others? Gibb (1968) suggests that he

- expresses directly to another whatever he is presently experiencing;
- communicates without distorting his messages;
- listens to others without distorting their messages;
- reveals his true motivation in the process of communicating his messages;

- is spontaneous and free in his communications with others and doesn't resort to habitual and/or planned strategies;
- responds immediately to another's need or state instead of waiting for the "right" time or giving himself enough time to come up with the "right" response;
- manifests his vulnerabilities and, in general, the "stuff" of his inner life;
- lives and communicates in the here and now;
- strives for interdependence rather than dependence or counterdependence in his relationships with others;
- learns how to enjoy psychological closeness;
- is concrete in his communication; and
- is willing to commit himself to others.

The person Gibb describes here is not a "free spirit" who inflicts himself on others. Freedom from role is not the same as license. It means that you don't use façades to overprotect yourself or fool others.

Consider an example. Your small group has a "trainer," and you are a "trainee." But, since you are members of a "learning community," one of your goals is to move toward role-free interaction within the group. If the trainer emphasizes his authority, intimates that he has knowledge of the group and its processes that he is not sharing with the other members of the group, becomes defensive when challenged, discloses little about himself, and generally remains aloof from interactions, he is locked into role behavior rather than role-free behavior. Or if you, as trainee, fear the trainer but refuse to share your fears with him or your fellow group members, are afraid to initiate anything that is not previously sanctioned by the trainer, see the trainer's providing directionality as principally an exercise of authority, and generally act in a dependent or counterdependent way toward the trainer, you are locked into the role of "trainee." The fact that the trainer tries to remain role-free does not imply that the trainer has no more to offer than anyone else. It does mean that when he "models" effective group behavior he does so not as a paragon but as a person who wants to be a good group member. Although roles are a legitimate function of social interaction (we don't expect a waiter or waitress to be "role-free" when we are trying to get served in a restaurant), they are also often used as defenses. The teacher uses his role to keep students at bay; students use their role to excuse a host of self-defeating behaviors. In examining your interpersonal style, try to discover whether you misuse any of your social roles—that is, whether you use them to keep you from engaging in life more fully. Describe some roles that you think would contribute to a lack of genuineness in the group (such as "rescuer," "sympathizer," "critic," and so forth).

Spontaneity. The genuine person is spontaneous. Many of the behaviors suggested by Gibb are ways of being spontaneous. The high-level communicator, while being tactful (as part of his respect for others), is not constantly weighing what he says. He doesn't put a number of filters between his inner life and what he expresses to others. He is assertive in communicating—in revealing himself and in responding to others—without being unduly aggressive. If you don't move out to meet others spontaneously in your day-to-day life, to do so in this group will seem foreign to you. Therefore, in the group you have to learn that it's all right to be active, spontaneous, free, and assertive. In my experience with training groups, lack of assertiveness is a very common problem.

The spontaneous person is free but not impulsive. If he weighs what he says, he does so out of concern for the person with whom he is communicating and not for his own protection. He is not constantly looking for rules to guide him in his relationship with others (not even the "rules" laid down here; rules may initially help him, but he doesn't become their slave). The good communicator is like a basketball player. He does learn some basic rules, but, more importantly, he learns basic skills. The basketball player has laboriously learned a number of "moves" and uses them whenever they are called for. In the same way, a high-level communicator has many responses in his repertoire. His ability to call on a wide variety of responses allows him to be spontaneous.

Nondefensiveness. The genuine person is nondefensive. He has a feeling for his areas of strength and deficit in interpersonal living, and presumably he is trying to live more effectively all the time. When someone expresses a negative attitude toward him, he tries to understand what the other person is thinking and feeling. Consider the following example.

> *Group Member A:* I don't think I'm getting anything out of these sessions. I'm just the same as I was weeks ago when this started. And I don't see any big differences in my relationships with others outside the group. I should probably leave and not waste my time here anymore.
>
> *Group Member B:* I think you're the one wasting time. You don't really do anything here, so what can you expect?
>
> *Group Member C:* That's your decision.
>
> *Group Member D:* There's been no payoff for you here—just dreary work with no results. I wonder whether it might help to examine a bit more carefully just what has been happening with you.

Members B and C are both defensive, but D tries to understand and gives A an opportunity to get at the issue of responsibility in the training process. The genuine person is at home with himself (although not in a smug way) enough that he can allow himself to examine negative criticism honestly. In this case, D would be most likely to ask himself whether he himself is contributing to what A feels. The person who is always defending himself has a great deal of difficulty involving himself with others.

Consistency. The genuine person doesn't drown in his own discrepancies. For instance, he does not have one set of "notional" values (such as caring, love, involvement, availability) different from his "real" values (comfort, power, self-gratification, being parental). He does not think or feel one thing but say another. At the same time, he doesn't dump his thoughts and feelings on others without discretion.

> *Group Member E:* I want to know what you really think of me.
>
> *Group Member F:* I think you are lazy, and that accounts for your lack of involvement here. You want to become more effective interpersonally by magic.
>
> *Group Member G:* I almost get the feeling that you want me to be brutally honest—with the accent on the "brutally." Maybe we can take a look at what has or has not happened between you and me, and both learn something about ourselves.

F is literal-minded and blunt, whereas G sees E's request as a desire for greater contact and immediacy. G knows that direct, solid feedback is important, but he prefers that it take place in a way that enhances rather than retards communication. Low-level communicators go to extremes: they are either blunt or timid. Both kinds of behaviors are rationalized—the former as "frankness" and the latter as "tact." Tact springs from strength, not weakness. The high-level communicator knows when he is tempted to hold something back for his own sake (for example, because he fears the reaction of the other person) and when he is doing so for the other person's sake (the other group member is not yet ready to hear it—although he will have to hear it eventually). The phony person is filled with discrepancies: he feels things but doesn't express his feelings, he thinks things but doesn't say them, he says one thing but does another. He is inconsistent.

Self-sharing. The genuine person is an open person, capable of deep self-disclosure. Self-disclosure, as we have seen, is not an end in itself for him, but he feels free to reveal himself intimately when it is appropriate. He enjoys mutuality in self-disclosure. This means that in a human-relations-training group he does not take on the role of helper or force another into the role of helpee. Later on we will see that genuineness can also express itself in confrontation and immediacy.

Exercise 37: A Checklist on the Communication of Genuineness

It is impossible to "practice" genuineness in and by itself, for genuineness is a set of attitudes and behaviors that modify the entire communication process. However, you and your fellow group members can give one another feedback on the quality of genuineness communicated during your training sessions. In order to give concrete feedback, you should have clearly in mind the *behaviors* that constitute the communication of genuineness. Here is a checklist of these behaviors.

1. Is the communicator his natural self? Or does he take on certain roles (helper, helpee, sympathizer, rescuer, organizer)? If so, what are these roles? Is his language his own, or is it stilted?
2. Is the communicator spontaneous (yet tactful), or is there something rigid and planned about his interactions? Does he respond to and contact others in stereotyped ways?
3. Does he avoid defensiveness? How does he handle negative confrontation? Does he avoid attacking others?
4. Is the communicator open? Does he share himself willingly and appropriately? Does he initiate such disclosure on his own, or does he have to be prodded?
5. Does the communicator express what he thinks and feels, without putting a number of filters between himself and the other members of the group?
6. Is his communication behavior generally consistent, or is he one person at one time and another later? Is he consistent but still spontaneous?
7. Does the communicator play games? If so, what games? Does he try to control and manipulate others? If so, how does he do so?

If you tell a person that he is not genuine, you should be able to back up your statement with concrete behavioral examples. This checklist should help you do that.

RESPECT: A BEHAVIORAL APPROACH

Respect for others, like genuineness, can be treated as a moral quality. However, here we are interested in the kinds of behaviors generated by respect and the behaviors through which respect is communicated. Respect, like caring, is "more than good intentions and warm regards" (Mayeroff, 1971, p. 69). When someone is interacting with you, how do you know that he respects you? What behaviors indicate respect?

Respect is such a fundamental notion that it eludes definition. The word comes from the Latin root that includes the idea of "seeing" or "viewing." Respect is a particular way of viewing another person. Respect means prizing another person simply because he is a human being. It implies that being human has value in itself. This kind of basic respect obviously goes beyond respect based on another's achievements. Choosing to prize others simply because of their humanity is also a value. But a value is a value only to the degree that it is translated into some kind of action. Respect as it is discussed here is an attitude expressed behaviorally. Some values induce people to act; others make them refrain from acting. That is, values can be active or passive. For instance, suppose justice is a value for me. If it is a passive value, I do nothing to cause injustice to others. If it is an active value, I do things to ensure that justice is promoted: I am active in various civil-rights movements, I fight for equitable tax laws, and so on. Some values, then, merely set limits on our behavior, whereas others galvanize us into action.

In the case of respect, actions literally speak louder than words. You will probably never *say* to one of your fellow group members "I respect you because you are a human being," "I prize you," or "I respect you for engaging in self-disclosure; you are doing a good thing." Rather, respect is communicated by the way you orient yourself toward and work with the members of your group.

Being "for" the other. Respect (in the sense discussed here), like genuineness, is not innate. If you want to exercise respect, you have to work to develop it. This kind of respect affects your manner of approaching others. Your manner indicates that you are "for" the other simply because he is human. This is not a sentimental attitude.

Respect ultimately involves helping others place demands on themselves (for instance, your fellow group members may want to change certain interpersonal attitudes and/or behaviors). "Being for," then, refers to the other's basic humanity *and* to his potential to be *more fully* human. Respect is both gracious and tough-minded.

Attending. Attending is itself a way of showing respect. It says, behaviorally, "I am with you. I am committed to your interests. It's worth my time and effort to work with you." Failure to attend is generally seen by others as a lack of respect. It says "You are not worth my time. What you say isn't worth listening to. I'm not really committed to working with you."

Willingness to work with others. Your training group is basically a learning community, and respect demands a willingness to give yourself actively to the work of the group. It is hoped that you see working toward high-level communication as a value and not just a chore. If learning communication skills, reviewing your interpersonal style, establishing relationships with your fellow group members, and attempting (at least on a basic level) to change some of your interpersonal behaviors are worth the investment of your time and energy, then the work you do will be an indication of your respect.

Regard for the other as unique. The individuality of others is usually a value for the person who is living effectively. If you are committed to supporting your fellow group members in their uniqueness, you will help them develop the resources that make them unique. This does not mean that you are in the group as a "helper," or that your intention is to make others over in your own image and likeness.

Regard for the other's self-determination. Your basic attitude should be that the other members of your group do have the resources they need to live more effective interpersonal lives. These resources may be blocked in various ways (for instance, a person has learned from childhood to fear intimacy with others), or they may be just lying fallow (a person may never have been challenged to use his resources). The members of the learning community, as they interact with one another, help one another free these resources or cultivate them. The expectation is, however, that each member is self-determining. Ultimately, if a participant chooses to live a less effective interpersonal life than he can, his choice should be respected.

Assuming the other's good will. Respect calls for your acting on the assumption that your fellow group members want to work at living more effective interpersonal lives. You should act on this assumption in any given case until it proves false (for instance, a group member shirks the work of the group). Even when a group member acts in such a negative way, you might ask yourself whether you or any other member of the group has contributed to his negativistic behavior. You should not abandon group members too readily because they seem unmotivated.

Suspending critical judgment. At least in the early stages of the group, respect takes the form of suspending critical judgment of others. Rogers (1961, 1967), following Standal (1954), calls this kind of respect "unconditional positive regard." This attitude means communicating to the other "a deep and genuine caring for him as a person with potentialities, a caring uncontaminated by evaluations of his thoughts, feelings, or behaviors" (Rogers, 1967, p. 102). Consider the differences in the following responses.

> *Group Member A* (revealing something about his interpersonal style): I think I'm a very selfish person. I establish friendships with people who let me have my own way. When I'm thwarted, I pout or find ways to make reprisals. At times I put my own self-gratification ahead of the more serious interests of my friends. And yet I'm "charming" enough to get away with it all. So it goes. Ugh.
>
> *Group Member B:* I'll bet this kind of selfishness hasn't really satisfied you. Ultimately it's going to make you miserable, if it hasn't done so already.
>
> *Group Member C:* When you look at yourself quite honestly like this, you see a degree of self-centered behavior that you don't like at all.

C neither judges nor condones. He merely tries to communicate understanding to A. This does not mean that C is naïve. He realizes that some of A's behaviors must change, but he still respects the person who is the author of these behaviors. He gives A room to move, room to explore himself. He realizes that judgmental behavior is likely to cut such self-exploration short.

But respect, even at this early stage, is not completely unconditional, for respect includes the assumption that the members of the group are willing to pay the price necessary for interpersonal growth. Respect, then, places a demand on the other (a demand he wants to place on himself) to grow, even while it offers the encouragement, the support, and the understanding needed for growth. If I help you remain aware of the demands you want to place on yourself and do so in a positive way, I am not reneging on my view of you as self-determining.

Accurate empathy. One of the best ways of showing respect is by working to understand the other members of your group. This work requires skill and patience. I know that a person respects me if he spends time and energy in trying to understand me. All of the behaviors associated with the communication of accurate empathy, then, are behaviors indicating respect.

Warmth. Gazda (1973) sees warmth as the physical expression of understanding and caring, which is ordinarily communicated nonverbally through gestures, posture, tone of voice, touch, and facial expression. Warmth is only one way of showing respect. It is not the best way, and it can be easily misused. Initial warmth lends substance to my attitude of being "for" you, but group members, certainly in the beginning, should not expect from one another the kind of warmth that takes place between good friends. At this point you are not good friends, and the relationships you develop need not be those of close friendship. For you to become a warmth machine, cranking out "unconditional positive regard" continually, would actually indicate lack of respect. Such warmth imparts an "oh-that's-all-right" attitude antithetical to the kind of interpersonal responsibility group members are trying to develop.

> *Group Member D:* I'm too easy with myself. As you can see, I've let myself go physically. I'm sort of fat. I don't read anymore and therefore provide little interest in conversations. And lately, I've found myself fighting with people—the way I've been doing here.
>
> *Group Member E:* First of all, Bill, most men your age let themselves go a little. You don't look so bad to me. Don't be harder on yourself than you should be.
>
> *Group Member F:* Things aren't right physically, intellectually, or interpersonally—and you don't seem to like it.

Member E translates warmth into an attempt to rescue D; he goes so far as to suggest that D lower his standards. On the other hand, F expresses appropriate warmth (assuming his tone is nonjudgmental) through accurate empathy.

Reinforcement. A climate of support arises partly from the encouragement group members show one another. This does not mean that you must become parental and reward the "good deeds" of your fellows. However, when a person succeeds in placing demands on himself to grow, recognition of his determination is quite reinforcing.

> *Group Member G:* I find myself taking more time to listen to others—not only here but outside as well. This has lessened my tension considerably. I didn't realize that such a simple thing could bring such improvement.
>
> *Group Member H:* If you add to that increased efforts to communicate understanding, I'll bet the picture will change even more dramatically.
>
> *Group Member I:* It sounds as if this makes you feel good about yourself!

While H's parental statement ignores what G has accomplished, I's accurate empathy recognizes and thus reinforces what G has done.

> *Group Member J:* I've got some bad news to report. Even though I talked a lot about controlling my temper in the group last week, at least three times I ended up blowing my stack with my family. I don't know whether I expected a dramatic change or what, but it was the same old me at home.
>
> *Group Member K:* I don't think you have to feel so bad. You can't expect to change your style all at once.
>
> *Group Member L:* You sound kind of disappointed in yourself. You want to change, but maybe it's not that easy.

Member K might well be reinforcing J's self-defeating pattern. L, on the other hand, tries to pick up J's feelings about himself; he is not judgmental, but neither does he condone J's behavior. Why should he, since J himself doesn't like this behavior?

Genuineness as respect. Being genuine in your relationships with others is a way of showing respect. Therefore, the behaviors that indicate genuineness, listed in the previous section, are also behavioral ways of indicating your respect for your fellows.

Exercise 38: A Checklist on the Communication of Respect

As with genuineness, it is impossible to "practice" respect in and by itself. Respect refers to a set of attitudes communicated to others and to behaviors embedded in the communication process. However, although you cannot practice respect separately, you can give one another feedback on whether your behaviors convey respect. Use the following criteria in giving one another feedback on the communication of respect.

1. Does the communicator seem to be "for" the other in a nonsentimental, caring way?

2. Is he *working* at communicating with the other members of the group?
3. Is the communicator speaking to the other as a unique individual and not just as an object or role-player?
4. Does the communicator avoid being judgmental?
5. Does the communicator use accurate empathy frequently and effectively?
6. Does the communicator give the other feedback on the other's *resources* (and not just on his deficits)?
7. Is the communicator appropriately warm? Does he avoid coldness, aloofness, "canned" warmth, inappropriately intimate warmth?
8. Does the communicator attend effectively? Does he use his body to make himself present to others?
9. Does the communicator avoid statements or behaviors that might indicate that he is trying to manipulate or exploit others?
10. Does the communicator find ways of reinforcing others for what they do well?

If you tell a person that he is not showing respect, you should be able to back up what you mean using concrete behavioral examples. This checklist should help you do that.

THE SKILLS OF RESPONDING: D-, M-, AND B-TYPES

As I have indicated before, rarely do we see pure "types" in training groups. However, since most of us have periods when we emphasize one kind of behavior over another, it may be helpful to see how each type uses responding skills (or fails to use them) in the group.

The D-Responder. The person with D-needs is so preoccupied with his needs and agenda that he finds it difficult to attend and to listen in an uncontaminated way. Attempts at accurate empathy, if made at all, either are very mechanical or miss the mark. When the D-person does respond, it is often with a presentation of his own needs. For instance, he may respond to self-disclosure with his own unrelated self-disclosure. The result is that such a person finds dialogue extremely difficult. The presentation of his own needs and agenda most frequently ends up as some kind of monologue. The passive D-person will withdraw from the interaction and not respond at all, not even nonverbally. The active or agitated D-person does not really attend in the way attending has been described here but rather is constantly on edge and overly present. Response from such a person often takes the form of irresponsible confrontation or advice-giving. The D-person locks himself into roles (the Timid Soul, the Angry Battler, the Needy Trainee, and such) and attempts to lock others into roles also (the Strong Person, the All-Knowing Trainer). The D-person often senses that he or she is more a helpee than a trainee but cannot deal with this realization directly.

In terms of what you have learned about response skills, can you fill this picture out further?

The M-Responder. The M-responder learns the technology of accurate empathy quickly enough, but he makes relatively little effort to use it with any consistency. This skill does not become second nature, rather, it has a mechanical or "exercise" quality about it. Accurate empathy, as a spontaneous skill, makes for greater intensity of interpersonal living, and the M-person is not sure whether he wants to live more intensively. Self-disclosure for the M-person is often a story to be told rather than the beginning of mutuality and dialogue; therefore he neither responds to the self-disclosure of others instinctively with accurate empathy nor expects to be responded to with accurate empathy. Such a person settles into the role of trainee, often finding it comfortable enough, but does not "stretch" to become a member of a learning community–responsible for both his own learning and the learning of the other members of the group. One effect of the M-person's way of dealing interpersonally with the group is to keep the trainer locked into his trainer role. As long as the trainer is seen as "something special," his or her behavior may be admired but need not be imitated. The M-person attends well enough, but he does not translate good attending and good listening into effective responding. The M-person is not spontaneous. He does well enough as long as the trainer provides structure for practicing skills, but, once the structure is withdrawn, he doesn't tend to initiate interactions.

The B-Responder. The B-person commits himself seriously to the training process without becoming swallowed up by it or locking himself into the role of trainee. Such a person responds to others with empathic understanding spontaneously, naturally, and consistently without becoming a "specialist" in that skill. Indeed, skills are consistently subordinated to the process of establishing and developing relationships. The B-person responds humanly to others, not as a helper to helpees. Since mutuality and dialogue are important for a B-person, he both responds with accurate empathy and expects it from others. But he weaves it naturally into a variety of responses, using whatever skill is appropriate at the time. The B-person is a caring person in his life outside the group, and therefore respect and caring are not commodities artificially tacked to his interpersonal style in order to fulfill the requirements of the contract. The B-person is active and eager without being a fanatic, and his eagerness places demands on others to respond in kind.

CHAPTER 8: FURTHER READINGS

On genuineness and respect:

Carkhuff, R. R. *Helping and human relations* (Vols. 1 and 2). New York: Holt, Rinehart and Winston, 1969.

Rogers, C. R., & Truax, C. B. The therapeutic conditions antecedent to change: A theoretical overview. In C. R. Rogers (Ed.), *The therapeutic relationship and its impact.* Madison: The University of Wisconsin Press, 1967. Pp. 97–108.
Stern, E. M. Psychotherapy: Reverence for experience. *Journal of Existentialism,* 1966, *6,* 279–287.

On game-playing, control, and manipulation:

Berne, E. *Games people play.* New York: Grove Press, 1964.
James, M., & Jongeward, D. *Born to win.* Reading, Mass.: Addison-Wesley, 1971.
Shostrom, E. *Man, the manipulator.* New York: Bantam Books, 1968.

PART 4

Phase I, Part 2: The Skills of Challenging

In Phase I you were introduced, both cognitively and experientially, to two sets of skills: the skills of letting yourself be known (self-disclosure, concreteness, the expression of feeling) and the skills of responding to others (the communication of primary-level accurate empathy and the behavioral communication of respect and genuineness). These two sets of skills provide the foundation for a supportive climate in the laboratory. Through self-disclosure (appropriate, of course, in the sense of being goal-directed), you have risked yourself—let yourself be known—in preliminary ways. And, by responding humanly to your fellow learners, you have justified the risk they have taken in disclosing themselves.

The function of this second part of Phase I is to introduce you to three more interpersonal skills. The skills of challenging pertain to both one-to-one relationships and to group interactions.

I have tried to be precise in writing this section, for it deals with skills that can be very "strong medicine" in human interaction. Strong medicine can be misused and abused, but it can also have dramatic positive effects. All human-relations skills can contribute to the social-influence processes discussed in the introductory chapter of this book, but the skills of challenge have special potency in this regard. However, I would like to suggest here that, in human-relations training, the skills of challenge are mainly invitational; that is, through these skills group members invite one another to consider overlooked or undealt-with aspects of interpersonal style and group behavior. These skills are not ways in which one member imposes his will or values on another. However, if you reveal your interpersonal values to the group and suggest behaviors that you would like either to develop or to drop in order to improve your interpersonal communication and other interpersonal behavior, the group may well use the skills of challenging in order to help you place these demands on *yourself*. Plato long ago suggested that the unexamined life is not worth living. Group members can use the skills of challenging to stimulate one another to examine their interpersonal living in new ways.

Support *and* challenge characterize high-level human relationships. A friend is a person with whom I can be "at home" but who is not afraid to make demands on me. His respect ultimately takes the form of challenging me to live up to my values and commitments. Challenging skills, then, are important in deep relationships. However, they are also focal skills in the helping process (see Egan, 1975b), and therefore there is a danger that you and your fellow trainees, in learning and practicing these skills, will be tempted to play psychotherapy and turn one another into "helpers" and "helpees." But the goal of the group remains the same: to examine your interpersonal style through the process of establishing and developing relationships with one another and getting feedback from one another. Since solid relationships demand both support and challenge, use of the skills of challenging is certainly legitimate in the process of relationship building. Care must be taken, however, to exercise these skills in the spirit of *mutuality*—mutual care, mutual concern. If you engage in the process of establishing and developing high-level relationships, you and your fellow trainees will undoubtedly help one another, but your help will be a by-product of your mutuality and not a substitute for it.

A PRE-CHALLENGING SKILL: FEEDBACK ON STRENGTHS

As we shall see, there is some negative element in even the best kind of confrontation. For instance, if I see a strength or a resource in you that you are not using, the identification of the strength or resource is positive, but my telling you that you are not using it is negative. In order to place the negative elements of challenging in a more positive context, you can develop the skill of giving others feedback on the strengths and resources that they have *and use.* Everyone has resources. One of the functions of this lab is to put you and your fellow group members in touch with your interpersonal resources, both those you are using and those you are not using. Sometimes we have negative self-images because we fail to identify our resources.

Exercise 39: The Identification of Interpersonal Resources

The purpose of this exercise is to make you aware of your own and others' interpersonal resources, to enable you to give feedback to others on the positive aspects of their interpersonal style, and to learn to provide a positive context for the negative elements involved in the challenging process.

1. Identify two or three interpersonal strengths in yourself and in each of your fellow group members.
2. Make your descriptions of these strengths as concrete as possible. Describe the strengths in terms of behaviors, and indicate what impact these behaviors have on others.

   Example. Interpersonally, you are a strong person. By that I mean that you're active. You initiate dialogue with others. You risk yourself by disclosing yourself rather deeply. And yet you don't seem overly anxious when you do any of these things. All of that strikes me as strength. It has two effects on me. It makes me feel good, because you add a dimension of honesty to the group. It also makes me a bit nervous, because you challenge me. What you do invites me to risk myself.

3. Share what you think your own strengths are with your fellow group members. This can be done in round-robin fashion.
4. Give feedback to others on their strengths. This can be done in the full group or in a round robin. If you do it in the full group, avoid needless repetition. If a person has been told a couple of times in detail that he has such-and-such a strength, when your turn comes you can indicate briefly that you share the opinions of the others.
5. As you receive feedback, see if you can identify the *patterns* of your interpersonal strengths.
6. Do you see strengths in yourself that others don't see? What strengths do others see that you don't see in yourself?

When group members give one another feedback on their strengths, a great deal of warmth and sense of community is usually generated within the group. Sometimes, however—especially in labs that are unstructured—the process stops here. There is little or no challenge. A group that is too cozy is as unproductive as a group that specializes in irresponsible confrontation. Balance is essential. If you want to help another by telling him what he or she fails to do, you should also be able to express to that person what he or she does well. In giving feedback to others, try not to lose sight of the larger context. Confrontation can include accurate empathic understanding; it can also include recognition of what the other does well. If you specialize in telling others what they do well without ever challenging them, or if you specialize in challenging others without any recognition of what they do well, your contribution may well be discounted.

CHALLENGING SKILLS AND RELATIONSHIP-BUILDING: A CAUTION

All challenging skills assume some kind of solidarity between the person challenging and the person being challenged. Only those who understand deeply have the right to challenge. Therefore, some of the exercises that appear after the sections on advanced accurate empathy, confrontation, and immediacy may not be appropriate if the group members have not had enough time to build up relatively solid relationships. If that is the case, it is better to defer these exercises until they flow more naturally from the process of relationship-building. Obviously, the trainers and supervisors of the human-relations program will help you make these decisions. Since presumably the group has been working hard to forge itself into a learning community, the skills of challenging will now begin to be appropriate.

Chapter 9

Moving Deeper: Advanced Accurate Empathy (AE II)

If you are the beneficiary of challenging skills used caringly and effectively by others, you will gain something very important. You will come to see yourself as others see you; you will discover what impact you have (or fail to have) on others. If you use these skills well, you will help others come to a better understanding of themselves. Sometimes we hang onto self-defeating patterns of interpersonal behavior because we don't even realize that our interpersonal transactions are less than they can be. The ancient Greek saying was "Know thyself." This has always been a human goal. The skills described in this chapter provide a technology that enables you and your fellow group members to know yourselves better.

Let's apply some of these principles to advanced accurate empathy. As you recall, through AE I you communicate to others an understanding of them from *their* frame of reference. AE II (and the other skills of challenging) offer other (external, possibly more objective) frames of reference, so that the other is given an opportunity to see himself more objectively—to stand outside of himself, as it were, and get a different picture of himself. Let's turn now to just how this is done.

THE MANY FORMS OF ADVANCED ACCURATE EMPATHY

If you work at "being with" your fellow group members, you will soon begin to know them in deeper ways, for you are constantly revealing yourselves to one another—by what you say, by the way you say it, by nonverbal behaviors, by your silences, by the differential ways in which you involve yourselves with different members, and by what you don't do. As you get to know one another in deeper ways, you also begin to share this knowledge with one another. Many authors call this kind of sharing "feedback"; others object to this term because it is too machine-related, not human enough. Whatever we call it, the process is important. In order to provide high-level feedback, you must be deeply involved with your fellow group members, you must be perceptive (a good discriminator), and you must be able to translate your perceptions into language that others can understand and assimilate (you must be a good communicator). Good attending, listening, perception, and communication are at the heart of advanced accurate empathy.

Consider the following example.

> *Group Member A:* I don't know what's going on in this group. I think I try just as hard as everyone else, but I still feel out of community. I don't seem to fit. I don't form my own relationships as easily as the rest of you. It's probably my own fault, but my efforts go down the drain. I don't know what else I can do.

One possibility is to respond with primary-level accurate empathy:

> *Group Member B:* It's frustrating and depressing. You work as hard as anyone else, but you don't experience the same success.

Member B understands A from A's frame of reference. He deals with A's feelings and the experience and behavior underlying these feelings. However, let's assume that this is not the first time that A (and the other members of the group) have received a good deal of understanding. In this case, another response may be appropriate.

> *Group Member C:* It's depressing to put so much effort in and still feel that you're not getting anywhere. It sounds as if you're pretty down on yourself, almost feeling sorry for yourself.

From the context, from past interchanges, from A's manner and tone, C picks up something that A does not express overtly: he is feeling sorry for himself. AE II, then, goes beyond what is overtly expressed to what is implied. If C is accurate, however—if his timing is good, if what he says and the way he says it don't sound like an accusation—then his response (AE II) can provide his fellow participant with some feedback that may help him move out of a self-defeating spiral. Member A may respond to Member C's comment something like this:

> *Group Member A:* You know, I've begun to wallow in self-pity instead of trying to figure out what I'm doing wrong. I want to examine my style here as concretely as possible, and I need feedback from you people in order to do it.

In this case, C's AE II pays off. A doesn't feel put off or judged; instead, he gets in touch with himself in a new way and is helped to move forward. We can probably assume that C's remark has helped because C has made efforts to "be with" A since the beginning of the group.

Even when the high-level communicator sees the other person's world from the other's viewpoint (as C does here), he often sees it more clearly, more widely, more deeply, and more cogently than the other. He has the advantage of being outside the other. He sees not only the other's perspective or frame of reference but also the *implications* (for effective or ineffective interpersonal living) of this perspective. The communication of advanced accurate empathy is the high-level communicator's way of sharing his understanding of these implications with the other.

One way of looking at advanced accurate empathy is as a form of intuition. As Steiner (1974) suggests, we use intuition a great deal in our day-to-day interactions with others.

> Let us assume we meet a friend on the street, and our intuition tells us that she is sad, exhausted, and frustrated, and that she may rebuff a warm greeting. We don't *know* for sure, but we can estimate what she will do and act on our estimate. We may decide that there is a 60% chance that she'll reject our stroke and so we approach her with a preliminary, cautious smile, waiting for her response or feedback. . . . This is the effective use of intuition: because it isn't 100% reliable it needs to be modified through feedback [p. 123].

In a sense, then, we often have *hunches* with respect to what is going on inside people we know. When we communicate these hunches to others, if we do so skillfully and caringly, we often help them get in touch with feelings or issues that they don't see clearly. For instance,

> Well, Jeff, I know that not getting the job makes you feel disappointed. But you also sound a bit angry, or at least I hear some anger in your voice. It sounds as though you think that you've been treated somewhat unfairly.

Sharing this hunch with Jeff may make it easier for him to express and deal with his anger.

In a sense, the skill of advanced accurate empathy is the skill of making our intuitions and hunches more accurate or exact. (However, since this skill is difficult, Steiner suggests that we be *cautious* in communicating our hunches to others.) Below are listed several ways of helping you make your intuitions and hunches more accurate. In any case, your accuracy will increase to the degree that you get feedback from those with whom you communicate. Therefore, feedback from your fellow group members on the accuracy of your intuitions and hunches is a critical part of your training.

Expressing What Is Only Implied

The most basic form of advanced accurate empathy is to give expression to what the other person only implies as he communicates ideas to the group.

> *Group Member D:* I've gotten in touch with resources for relating here that I never realized I had. At least in any full way. I see that I'm caring, that I can talk

concretely, that I'm unafraid to reveal myself to others who give me half a chance. I'm not trying to blow my own horn. I'm just saying that these discoveries are important for me.
Group Member E: These resources are very real, and their discovery has been—well, exciting for you.
Group Member F: I hear a note of determination. Now that you've gotten in touch with these resources, you're going to make them a part of your interpersonal life. And that's even more exciting than doing well in the group.

F's AE II goes beyond E's AE I. There is a deeper message in what D has said than just satisfaction with his group experience. F perceives the implications of what D is saying and communicates his understanding to D. If F is accurate (and he bases his understanding, as we have already seen, on the total context of D's communication), his accuracy will help D get in touch with himself more objectively as a "man on the move" in an interpersonal sense.

Let's take a look at another example. In this example, Group Member G has been exploring her experiences, both inside and outside the group, of people who "don't give her a chance." As we take up the conversation, she is talking about someone she considers a very good friend.

Group Member G: I don't know why she acts the way she does. One day she'll be chattering away on the phone in a most engaging way. She's carefree and tells me all that's happening. She's great when she's in that mood. But, brother! At other times she's actually rude to me and moody as hell. And it seems so personal—I mean not just that she's in a bad mood generally, but that somehow it's directed at me. I'm usually a free, spontaneous spirit, but she really shuts me down when she's like that.

Group Member H: Her fickleness stops you dead in your tracks. Since her behavior seems to be more than just general moodiness—it's directed at you personally—I wonder whether, in her eyes anyway, you may do something to provoke such a reaction. I'm not sure whether you're intimating that it might be a question of two free spirits ruffling each other's feathers.

The implication of being the object of someone's anger is that one has done something (advertently or inadvertently) to provoke that anger. H's response tries to go below the surface a bit and open up an area for G to consider. H doesn't accuse or blame, but he does follow a lead that is implicit in G's statement. It may well be that H is also basing his response on the wider context of G's behavior in the training group.

Through advanced accurate empathy, you and your fellow group members place demands on one another to take a deeper look at interpersonal style. The rapport that has been created through the communication of genuineness, respect, and understanding has established—in social-influence terms—a power base, and now the group members use this power to influence one another to take a look at frames of reference other than their own. These demands are still based on accurate understanding of one another, but they go beyond surface understanding. The low-level communicator finds it difficult to make such demands on his fellow group members.

The Summary

You can also communicate advanced accurate empathy by bringing together and summarizing what the other has presented in bits and pieces during the group experience. This summarizing helps the other person focus on his interpersonal behavior in a new way. He sees himself from a different frame of reference.

> *Group Member B:* Let me see if I can put some of this together. Different things seem to keep you from making contact with different people here. George and Jim are too strong, too assertive, and their strength makes you keep your distance. On the other hand, you're afraid that Sarah might become dependent on you—for this is her tendency anyway. You and Mary haven't been able to find any "common ground." You want to establish a closer relationship with me, but you hesitate because you're not sure that I want any kind of relationship. I don't think that I'm misrepresenting what you yourself have said in your interactions here, but I'd like you to check me out on it.
>
> *Group Member A:* Oh boy! I really hadn't realized that in one way or another I was saying "No!" to everyone here. I'm so damned cautious. I do the same thing in everyday life. Everything has to be perfect before I'll enter a relationship. I'm prepared to take practically no risks.

All that B has done is to bring together data that A himself has produced in the group. Pulling it together in a summary way helps A identify a self-defeating interpersonal pattern. Now that he sees his style a bit more clearly, the possibility of changing his behavior increases. When a fellow group member talks in a rambling or confused way, sometimes a summary will help him refocus.

Identifying Themes

AE II includes the identification of behavioral and emotional themes as group members go about the business of exploring their interpersonal styles and establishing relationships. For instance, without saying so explicitly, a group member may intimate through what he reveals about himself and through the way he behaves in the group that he tends to be a dependent person.

> *Group Member B:* I've been thinking of some of the things you've said here, Ned, and trying to hook these up with the ways I see you acting here. You're hesitant to challenge anyone, especially the "strong" members in the group. You say things like "I've been quiet because other people haven't given me the kind of feedback I need." You've asked me outside the group for feedback on how you

are coming across. The message I tend to get out of much of this from you is "I'm not my own person. I depend on others quite a bit. I don't value myself; I wait to see how others value me first." Perhaps this is too strong. I'd like to find out how you're reacting to what I'm saying.
Ned: What you're saying is very painful for me to hear. It's even more painful because of the truth in it.

Let's take a look at another example.

Group Member D: Let's see if this makes any sense to you. On the one hand, you feel quite lonely. You even feel lonely and isolated in this group experience. On the other hand, I pick up cues here and there that you're somewhat reluctant to get close to others. You mentioned early in the group, for instance, that you hoped that people would take their time in making demands of you. It seems almost as if intimacy demands a price that you're not quite sure you're ready to pay.

Group Member C: Yes, and I seem to suffer either way.

Here again, D goes beyond what C has said explicitly. The thematic material pointed out by the listener may refer to feeling (such as failed enthusiasm, depression, success, joy, anxiety), to behavior (such as controlling others, caring, avoiding intimacy, cooperation, blaming others), to experiences (being a victim, being loved, being seduced, being feared, failing), or to combinations of these. This task demands a high degree of accuracy, tact, and initiative. If you try to force even highly accurate advanced empathy down the throat of one of your fellow group members, he will probably balk. On the other hand, if the communication of such understanding is tactful, you can help the other person see his interpersonal behavior in a new light. As you read some of the communications of accurate empathy in this section, they may well seem premature to you; but remember that they are taken out of context. The assumption is that rapport has been established and that the relationship is ready for this kind of interchange.

Helping Another Draw Conclusions from Premises

Still another way of conceptualizing advanced accurate empathy is to help another person draw his own conclusions from the premises that he himself lays down.

Group Member E: I know I'm quiet here, but I don't believe that's any reason for people to pick on me, to make me feel like a second-class citizen. I've got resources. I can say a lot of the things that others say, but they move in first.

Group Member F: The logic of what you're saying seems to be: one, I don't want to appear to be a second-class citizen; two, I have what it takes to be an active member; three, I'm going to begin to take my rightful place in this group. I'm not sure whether I'm saying too much.

"Drawing a conclusion" will often deal with how the other person wants to behave within the group (or in interpersonal relations generally.) If the "conclusion" you help your fellow group member draw is not in his premises, your empathy will be seen as an attempt to make him behave in ways that are acceptable to you. Again, accuracy is extremely important, as are tact and timing.

From the Less to the More

In summary, one way to look at AE II is to see it as helping and being helped to move from the less to the more. For instance, if one of the group members speaks ambiguously, guardedly, and/or indirectly about an issue that is important for him and important for the group, you, in contrast, can speak clearly, openly, and directly. Let's say that one of the group members has in very indirect ways touched on the issue of interpersonal power within the group. You can find out whether he or the other members of the group want to face these issues more directly.

> *You:* George, several times you have alluded to the "powerful" people here. You've hinted, at least it seems to me, that a couple of us make the rules here, even though there is a general contract. You've talked to Susan about her "weakness." I wonder whether you think it might be in everyone's best interests to find out how we see one another in terms of power and how all of this affects our group. I guess I'd like to see us air this as directly and concretely as possible.

One way of looking at advanced accurate empathy is to ask yourself "What is the *underlying* issue here?" Or "What is the *cloudy* issue that should be made clear?" Through AE II, what is said confusedly by a group member is stated clearly by you; what is said half-heartedly is stated cogently; what is said vaguely is stated specifically and concretely; what is presented at a superficial level is re-presented at a deeper level. In a sense, through AE II you *interpret* what another person says or does, but your interpretations are based on the other person's behavior, both verbal and nonverbal—not on psychodynamics or on abstract psychological theories.

THE IMPORTANCE OF MUTUALITY IN COMMUNICATING AE II

If the communication of advanced accurate empathy is a one-way street, or if you or other members of the group specialize in this (or any) particular skill, then the group is in trouble. The specialist in advanced accurate empathy can't help sounding like a psychologist. He will be seen in the group as a "helper," and even though the group process may well be extremely helpful, you are not there *as a helper.* You are in the group as a peer, and you and your fellows are pursuing group goals together. In the examples given so far, AE II has seemed like a one-way process. It need not be. This does not mean that in any given interchange there must be an exchange of advanced accurate empathy but, rather, that *all* participants should learn this skill and use it whenever it seems beneficial to do so. Unless there is mutuality in AE II, the group will give itself over to helping behavior instead of relationship-building. Self-conscious helping, as Gibb (1971) points out, is not helpful in human-relations-training groups; but mutuality is. Gibb distinguishes "orientations that help" from "orientations that hinder":

1. Reciprocal trust (including confidence, warmth, and acceptance) stimulates, whereas mutual distrust (including fear, punitiveness, and defensiveness) hinders.
2. Cooperative learning (including inquiry, exploration, wonder, quest) stimulates, whereas indoctrinating (including advice-giving) hinders.
3. Mutual growth (including becoming, actualizing, fulfilling) stimulates, whereas evaluating (including fixing, correcting, providing a remedy) hinders.
4. Reciprocal openness (including spontaneity, candor, and honesty) stimulates, whereas one-way helping and communication strategies and tactics (including planning *for* the other, maneuvering, and gamesmanship) hinder.
5. *Shared* problem-solving (defining, producing alternatives, and testing) stimulates, whereas parental-like guidance (including setting oneself up as model, demonstrating, and information-giving) hinders.
6. Encouraging autonomy (freedom, interdependency, equality) stimulates, whereas control (including coaching, molding, and steering) hinders.
7. Experimentation (including play, innovation, and trying new behaviors) stimulates, whereas standard-setting (including patterning) hinders.

In groups in which mutuality is the focus (including human-relations-training groups such as the one in which you are participating), help is not very helpful unless it is quite mutual. Study the following example.

Group Member A: Peter, you and I talk from time to time in here, but there is a very tentative quality about it. I see you as having a kind of approach-avoidance thing with me. You seem on the verge of opening yourself up to me often, but then you withdraw—or we both withdraw, and nothing happens again for a couple of weeks.

Peter: I was wondering when we would get around to discussing this openly. I *am* hesitant in deepening a relationship with you, even though I want to. From my point of view, when I do make overtures to you, when it begins to be a two-way street and not just something controlled by you, something changes—your spontaneity. You begin talking to me very formally. A kind of heaviness enters into our interactions, and I'm put off. We must want to get closer, because we keep repeating the same pattern, even though it doesn't work.

Group Member A: So I give the impression that everything is all right as long as

I'm in charge—but if not we both lose our spontaneity, and our relationship becomes a heavy, burdensome thing. Yet you seem hesitant to tell me what you see happening *when* it's happening.

Peter: You're darn right. I sit here with all these thoughts and feelings inside me, and I just close down. Now that this is out in the open, I'm counting on myself, you, and the others here to face up to it.

Both of these members use one or another form of advanced accurate empathy, but they do so not as helpers but as two people who are trying to deepen their relationship. They are not playing psychologist with each other; there is no attempt to explore hypothetical psychodynamics. They are trying to understand their own and each other's behavior. Mutuality should always characterize whatever "helping" goes on in your training group, which is a learning community in which all members are contributors. Advanced accurate empathy is impossible for you and your fellow group members unless you are exploring one another's lives more deeply and beginning to see interpersonal themes emerge as you interact with one another.

EXERCISES IN ADVANCED ACCURATE EMPATHY

The following exercises are designed to help you examine your own interpersonal style and the styles of your fellow group members in greater depth. Remember that AE II is one of those skills that can be called "strong medicine." Use it to add depth to your interactions with others, but don't highlight the confrontational aspect of it. Remember also that you are trying to understand and *involve* yourself with your fellow group members. AE II should promote mutuality rather than the problem-centered atmosphere of counseling.

Exercise 40: Advanced Accurate Empathy: An Exercise in Self-Exploration

One way of making sure that you are careful in using confrontational skills is to use them on yourself first. The purpose of this exercise is to make you think about some dimensions of your interpersonal style and behavior at two different levels (roughly corresponding to AE I and AE II). Self-understanding and being in touch with your own feelings and emotions should (at least logically) precede being in touch with deeper dimensions of the interactional style of others.

Directions

1. Read the examples given below.
2. Choose some situation or issue or relationship having to do with your

interpersonal style that you would like to take a deeper look into. Choose something that you will be willing to share with your fellow group members.

3. Briefly describe the issue (as is done in the examples).
4. As in the examples below,
   a. Write a statement that reflects a primary-level accurate empathic understanding of the issue you have chosen. This statement should reflect understanding of both your feelings and the behaviors/experiences underlying these feelings.
   b. Write a statement that reflects your advanced accurate empathic understanding of this issue. This understanding should go deeper into the issue.

Example 1

Issue. I'm concerned about the quality of my "being with" others in interpersonal and social situations.

a. I enjoy being with people. I meet people easily, and I'm generally well received and well liked. I make others feel at home. I'm outgoing and, to a degree, uninhibited when I'm with others—I'm humorous, I try to understand the world of others, I show an interest in what they are doing. I also try to be careful with others; that is, I try not to be "too much" (AE I).

b. When I'm with people, even though I am outgoing, I'm not "all there." I don't tend to share myself deeply with others. Therefore there is something almost superficial (perhaps this is too strong a word) about my "being with" others. In my deepest moments, I am alone with myself. Perhaps I haven't learned to share my deeper self with anyone. I may even be afraid to do so (AE II).

Example 2

Issue. I'm concerned with the way in which I participate in the human-relations-training group.

a. I'm an active group member. I enjoy trying to establish relationships with others. I'm open about myself. I encourage others to talk about themselves and spend a fair amount of time just understanding others. Others see me as a good group member, and I never get challenged on the quality of my participation. I am technically good at most human-relations-training skills, including confrontation. Most people in the group (if not all) think that I would make a fine trainer. I am very comfortable in the group (AE I).

b. On deeper reflection, there are ways in which I am dissatisfied with my participation in the group. First of all, I think that I exercise a great deal of control over what happens in the group. I share myself, but not too deeply. I see unwritten laws being made with respect to safety and security (at the price of more intensive interaction) within the group. I am limiting the intensity of the group by not even raising the issue of intensity—my own and that of others. I'm sliding through this group experience and getting a great deal of reinforcement in doing so. I feel phony, and yet I'm reluctant to violate my own comfort, especially when others see me as good. All of this smacks of some kind of manipulation, and I feel the edge of guilt. I often ask myself whether this is the way I want to live my life (AE II).

Now write out four combinations of AE-I and AE-II statements on interpersonal and group issues that concern you. Be as concrete as you can in both AE-I and AE-II statements.

Share your most significant statements with your fellow group members. Get their feedback with respect to "cutting deeper" in your AE-II statements. Are you getting a feeling for what it means to "go deeper"? Give similar feedback to your fellows.

Exercise 41: Advanced Accurate Empathy: Understanding Others More Deeply

Before trying to use advanced accurate empathy "on the spot" in your group, you can, through this exercise, prepare yourself at a leisurely pace for such empathy. This exercise should help you look before you leap into advanced accurate empathy.

Directions

This exercise is the same as the previous one, except that your attention is now directed toward your fellow group members instead of yourself.

1. Read the examples given below.
2. Consider each of your fellow group members one at a time. Choose some dimension of each one's interpersonal style that you would like to explore at both the AE-I and the AE-II level. Remember, you are trying to *understand* the person more deeply, not "psych him out."
3. Briefly describe the issue for each (see the examples).
4. As in the examples below,
   a. For each, write a statement that reflects a primary-level accurate empathic understanding of the issue you have chosen (feelings and content). Write the statement as if you were speaking directly to the person.
   b. For each, write a second statement that reflects your advanced accurate empathic understanding of this issue. Write the statement as if you were speaking directly to the person.

Example 1

Issue. John is both satisfied and dissatisfied with the strengths in interpersonal relating he manifests in the group.

a. John, you come across in the group as very self-assured. Most of the group members seem to enjoy interacting with you, and they do so frequently. You are understanding. You reveal how you feel about each of the group members without "dumping" your emotions on them. At one level, you seem to enjoy your position in the group. You get a great deal of respect and even admiration, and you find something satisfying in this (AE I).

b. John, even though you share yourself a great deal, by telling others how you are reacting to them as the group moves along, you don't speak much about your own interpersonal style. It may be that you are hesitant to do so, or that others box you in by putting you on a pedestal. Whatever the case, I sometimes see you wince slightly when you get positive feedback. I'm beginning to suspect that you feel you aren't allowed to share your vulnerabilities here, and that you're beginning to resent it (AE II).

Example 2

Issue. Jane is honest about her interpersonal deficiencies, and this honesty relates to behavioral change or lack of it in various ways.

a. Jane, you share your weaknesses and your fears with us quite freely. You've even said that you see yourself closer to the treatment end of the treatment-training continuum. Your honesty gives you a great deal of freedom, which you seem to appreciate. You also get a great deal of respect for being honest, which seems to buoy you up. Despite your admitted failings, you're accepted here, and I believe you feel this acceptance deeply (AE I).

b. Jane, as I look at you more closely, I think I see in you some disappointment in yourself. It's as if some of us here have said "Honesty is enough, Jane"—and you are longing for some concrete change in your behavior. Assertiveness in admitting your faults is not enough. As you've said, you are not assertive in making legitimate demands on others. Because you don't do that, you seem at times to feel useless. You may even wish that some of us would make more demands on you. I think you may fear that you will leave this group not much different from when you entered it (AE II).

Now write out combinations of AE-I and AE-II statements for each of your fellow group members. Try to be as concrete as possible.

Using a round robin or some other technique suggested by the trainer, share your statements with your fellow group members. Get feedback on the accuracy of your statements. Are they useful statements, or do they tend to be overly psychological?

CHAPTER 9: FURTHER READINGS

Bullmer, K. *The art of empathy: A manual for improving accuracy of interpersonal perception.* New York: Human Sciences Press, 1975.

Carkhuff, R. R. Laying a base for initiating. Chapter 4 in *The art of helping: Trainer's guide.* Amherst, Mass.: Human Resource Development Press, 1975. Pp. 86-TG 122.

Rogers, C. R. Empathic: An unappreciated way of being. *Counseling Psychologist,* 1975, *5* (2), 2-10.

Chapter 10

Confrontation

"Confrontation" is a word that inspires fear in many people, and perhaps rightly so, for they have seen themselves or others devastated by irresponsible interpersonal attack. At any rate, confrontation in human-relations-training groups is certainly a controversial issue, and both helpers and human-relations-training specialists continue to argue its pros and cons, its effectiveness and its risks. Confrontation has been the topic of a certain minimal amount of research (see Berenson & Mitchell, 1974, for a summary), and most of this research is related to helping rather than to human-relations training. Lieberman, Yalom, and Miles (1973), in studying a variety of approaches to sensitivity-training and encounter groups, found that confrontation in groups with low structure and low support led to a number of "casualties." To add to the confusion, there is no standard definition of confrontation and no agreement in the literature on what results it is supposed to have. Berenson and Mitchell (1974, p. 111) describe five different kinds of confrontation, but they warn the reader so stridently about the possible abuse of confrontation that they seem almost to discourage its use altogether. Their thesis is that relatively few people are living effectively enough and are skilled enough to merit the "right" to confront others. Be that as it may, interpersonal life abounds in confrontations—for better or for worse. Confrontation has been a fact of everyday life throughout history. Kanter (1972, pp. 14ff, 37ff), for instance, in a study of different kinds of communes, found that both confrontation and self-criticism have characterized a wide variety of successful utopian communities. It is senseless, then, to suppose that confrontations can be eliminated from interpersonal transactions.

Given this confusion, you may well ask yourself: should I confront the members of my group or not? The answer is that it depends. Confrontation, if used skillfully by a caring person (and Berenson and Mitchell certainly recognize this), can serve the interests of both parties in an interpersonal transaction. No more definitive answer can be given, however, without a more concrete explanation and description of confrontation—its nature, its goals, its potency, its limitations, and the conditions under which it can be used constructively. The purpose of this section, then, is to help you understand the nature and the technique of confrontation so that you can decide what place it should have in your interpersonal repertoire. In these pages I would like to give confrontation a "better name" and describe a process that, if carefully handled, can enrich interpersonal transactions and relationships.

THE ANATOMY OF CONFRONTATION

Inadvertent Confrontation

The process to be described in the following pages is *intentional* confrontation, or confrontation as a specific skill. However, confrontation is also "in the eye of the beholder"—that is, any transaction that is perceived by a person as confrontational. In this sense, even accurate empathic understanding can be seen as confrontation. Let me explain. If you attend carefully to another person, listen to what he has to say (both verbally and nonverbally), and communicate to him in a nonjudgmental way understanding of his feelings, experiences, and behaviors, you do two things. First, your communication is an act of intimacy, at least broadly defined. And, if the other person is fearful of intimacy, your getting close to him through an act of accurate empathy may well frighten him—that is, confront him with his own fears. Second, accurate empathy, if it is really on the mark, has a way of encouraging the other person to disclose or explore himself even further. In other words, there is a social-influence dimension even to primary-level accurate empathy. High-level accurate empathy is both supportive and demanding. Therefore, if for one reason or another the other person is reluctant to take a deeper look at himself, accurate empathy may appear confrontational to him, even if confrontation is not the intention of the speaker.

Advanced accurate empathy, insofar as it "digs deeper" and expresses what is only implied, is confrontational in itself, for it invites the other person to look at himself from a different perspective or in a new way, thus challenging him.

Confrontation as Invitation

"Confrontation," as used here, is anything that you do that invites a person to examine his behavior and its consequences more carefully. Confrontation as invitation may be indirect. For instance, if you act patiently and caringly with a group member with whom I have been quite impatient, I may well experience your behavior as confrontational in that it invites me to assess my own (in this case, ineffective) behavior. In these pages we will stress the invitational character of confrontation and focus on direct verbal confrontations. Let's look at an example of direct verbal confrontation.

> *Group Member A:* Mark, I wonder whether you are aware that, with some frequency, you change the topic or focus of a conversation when you respond to others. At least it strikes me that way. For instance, a little earlier Jim was talking about his fear of being controlled by people here whom he sees as stronger than himself. You responded to him, I think, with understanding; but somehow, as you talked about your own fears, *you* became the focus. And I'm not sure that Jim got a chance to explore his own fears. I hope I'm not distorting what actually happened.
>
> *Mark:* Now that you mention it, that's just what happened. I'm not too aware that I do this as part of my style, but I'd like some help in monitoring that. I wonder whether anyone else here has noticed that.

Member A invites Mark to take a look at a possibly nonfacilitative dimension of his interpersonal style, and Mark responds by accepting his invitation. Maintaining the invitational character of confrontation is important. Otherwise, confrontation degenerates into parental (rather than adult-to-adult), accusatory, punitive, and/or recriminatory behavior, and the mutuality necessary for good interpersonal transactions is lost. It is usually not necessary to hit someone in the head in order to get him to examine his behavior. The invitational character of confrontation is part of the tentativeness that should characterize the use of all challenging skills.

Types of Confrontation

Berenson and Mitchell (1974) describe five types of confrontation. Let's examine these as a way of getting into *how* and *what* to confront.

1. Didactic confrontation. This type of confrontation deals basically with information or misinformation. If you give one of your fellow group members information he doesn't have, or correct misinformation he has about relatively objective aspects of the world, including the "world" of the training group, then you are dealing in didactic confrontation. This type, too, can be invitational. It need not focus on the "ignorance" of the other.

> *Group Member C:* Rita, there's one aspect of the group contract under which we're working that I'm not sure that you're aware of. At least I'd like to check it out. You refer fairly frequently to what happens in your interpersonal life outside the group. For instance, you've talked a great deal about dependency problems with your mother and father. But often enough we deal with them as problems "out there," and not much is done to relate what you say to your relationships within the group. And it turns into a kind of counseling session.

This isn't bad in itself, but our contract asks us to relate there-and-then happenings to the here and now of our own group. Does this make any sense to you?

Rita: Now that you bring it up, I remember reading that in our contract, and I know I've let it slide. Yes, and I do end up in a counseling session—all of you become my counselors. I think two things are happening. First, I didn't read that part of the contract carefully enough. Second, my problems with my parents are sometimes so pressing on my mind when I come in here that I guess I'm just looking for help.

Member C is trying to find out whether Rita has some mistaken notion about the contract. His "invitation," therefore, centers around an information/misinformation issue. His hypothesis is that Rita's lack of information is getting in the way of her effective involvement with her fellow group members; and, as Rita's nondefensive reply indicates, to a degree he is correct. In this case a didactic confrontation helps clear the air and give sharper focus to Rita's problem of immediacy of involvement in the group.

2. Experiential confrontation. As you involve yourself more deeply with your fellow group members, you will notice that at times your experience differs from that of another group member: the other *experiences* himself, what is happening in the group, another group member, or you *differently* from the way you do. While avoiding making your own experience normative, you can still invite the other to examine these differences with you and the others. Let's take a look at a few examples.

Group Member E: Joan, you say that you're unattractive, and yet I know that you get asked out a lot. And, if I'm not mistaken, I see that people often react to you here with ways of saying "I like you." I can't seem to put this together with your being "unattractive."

Joan: Okay. What you say is true, and it helps me clarify what I mean. First of all, I'm certainly no raving beauty, and when others find me attractive, I think they mean personality—the fact that I'm a rather caring person and things like that. I think at times I wish I were more physically attractive, though I hate to admit it. But, more important, most of the time I *feel* unattractive. And sometimes I feel *most* unattractive at the very moment people are telling me directly or indirectly that they find me attractive.

Since Member E's experience of Joan seems to be different from Joan's experience of herself, he invites Joan to explore that difference in the group. Joan's self-exploration clarifies the issue greatly.

Group Member F: Jim, I'll bet I'm not far wrong if I say that you see yourself as a person who appreciates humor and is fairly witty. I've often seen you be genuinely witty, but sometimes I think what you see as wit I see as perhaps

> cynicism or even, at times, sarcasm. It could be that I'm oversensitive, and I think it would be good to get feedback from others here.
> *Jim:* Yeah, I think I'd like to hear what others have to say.

Here Jim does not say how he experiences himself, but he is willing to hear how others experience him.

> *Group Member G:* Nancy, you say that you see Cheryl as "pushy," making demands of you. I might have another perspective. I guess I see Cheryl as gutsy. She's committed to the goals of this group, and she does make demands of you, me, herself, all of us. But I see these demands as falling within the scope of the group contract. Cheryl makes me uncomfortable at times, but I need to be made uncomfortable.

Nancy is invited to review her experiencing of Cheryl. This does not necessarily mean that G's view of Cheryl is correct. It *is* different from Nancy's, however, and worth exploring.

What are the sources of differences in experiencing? Discrepancies, distortions, evasions, games, tricks, and smoke screens, to name a few—and all of us are guilty of some of these at one time or another.

a. Discrepancies. In all of us there are various discrepancies—between what we think and feel and what we say, what we say and what we do, our views of ourselves and others' views of us, what we are and what we wish to be, and what we really are and what we experience ourselves to be. And, of course, discrepancies are common between our verbal and nonverbal expressions of ourselves. For instance:

- I'm confused and angry, but I say that I feel fine.
- I see myself as witty, whereas others see me as biting.
- I experience myself as ugly, when in reality my looks are somewhat above average.
- I say "yes" with my words, but my body language says "no."
- I say that I'm interested in others, but I don't attend to them or try to understand them.
- I say that I want to improve my interpersonal style, but I'm lazy and listless and don't put much effort into this group experience.

In this lab experience we are interested principally in discrepancies that affect interpersonal style and group participation.

b. Distortions. If we can't face things as they really are, we tend to distort them. The way we see the world, including our interpersonal world, is often an indication of our needs rather than a true picture of what the world is like. For instance:

- I'm afraid of you and therefore I see you as aloof, although in reality you are a caring person.

- I see the group leader in some kind of divine role, and therefore I make unwarranted demands of him.
- I see my own stubbornness as commitment.
- I see your talking about yourself in generalities as adequate self-disclosure because I am afraid of disclosing myself.

One way group members can help one another is to suggest to one another alternative frames of reference for viewing self, others, and interpersonal living. For instance, it is confronting to suggest that

- living more intensively in the group can be seen as challenge rather than as just pain;
- what seems like heroic forbearance in the face of painful learnings about oneself might be just self-pity;
- a person might be something of a seducer rather than merely a victim;
- a person might be afraid to act rather than unable to act;
- caring for another might be smothering rather than nurturing; or
- intimacy is rewarding rather than just demanding.

Seeing alternative frames of reference helps us break out of self-defeating views of self, others, and the world of relating.

c. Games, tricks, and smoke screens. If I am comfortable with my delusions and profit by them, I will obviously try to keep them. If I'm rewarded for playing games—that is, if I get others to meet my needs by playing games—then I will continue a game approach to life (see Berne, 1964; Harris, 1969; James & Jongeward, 1971). For instance, I'll play "Yes, but . . . ." That is, I get others to give me feedback on my interpersonal style and then proceed to show how invalid that feedback is. Or I make myself appear helpless and needy in the group (or to my friends outside the group), and when they come to my aid I get angry at them for treating me like a child. Or I seduce others in one way or another and then become indignant when they accept my covert invitations. The number of games we can play in order to avoid intimacy and other forms of effective interpersonal living is seemingly endless. When we are fearful of changing elements of our interpersonal behavior, we often attempt to lay down smoke screens to hide from one another the ways in which we fail to seize life. We use communication in order *not* to communicate (see Beier, 1966).

The best defense against game-playing in your training group is to create an atmosphere in which it is almost impossible to play games. Effective group participants don't get "hooked" into games. For instance, if you refuse to play the role of "helper" in your group, others are prevented from playing the "Yes, but . . ." game. However, if someone in the group does try to play any of the innumerable games available, he should be challenged in a caring and responsible way.

> *Group Member A:* Maureen, you are the one in here that I can really talk to. You listen to me, and I'm sure you care about me. When I need feedback, I know I'm going to get it straight from you. You seem to know me so well. Your feedback is so accurate.
>
> *Maureen:* I appreciate your trust very much. But I'm also uncomfortable being singled out like this. I'm not sure whether you're implying that you can't develop other relationships in here similar to ours.

Member A involves himself in a game of "pairing," linking himself closely to another group member in seeming opposition to another or others. Maureen, however, doesn't let herself get hooked into "pairing." She challenges A to begin thinking about developing other strong relationships in the group.

3. *Strength confrontation.* Confrontation of strengths means pointing out to another the strengths, assets, and resources he is failing to use or is not using fully.

> *Group Member B:* Rick, I'd like to make a comment on the *quality* of your interactions here. The times you've interacted with me, I've really listened, because you are totally present. You attend, listen, understand very well. I guess I'm not sure why you're not available to more people here in that way consistently. Your interactions with me are so rewarding that, when you "retire" for an extended period, I miss you. But, overall, a certain lack of *quantity* of interactions on your part seems in the long run to affect even the quality of your presence here. I'm just not sure why you tend to hold back.

Here B places a demand on Rick that he deploy unused resources. If, as many social scientists have told us, we tend to use less than a quarter of our human potential (including interpersonal potential) as we make our way through life, then strength confrontations should be the most widely used kind of confrontation. Such confrontations are obviously *experiential,* for they must be realistic–that is, based on observation of real strengths, assets, and resources that have been developed or are not being used. In their research studies, Berenson and Mitchell (1974) have demonstrated that high-level communicators do indeed use strength confrontations more frequently than low-level communicators do, and they use them instinctively. In the group, then, you and your fellow group members are asked to demand the best from one another in terms of interpersonal transactions. There is a negative element even in strength confrontations, for it means pointing out possible strengths that are *not* being used. Overall, however, such confrontations are quite positive, for emphasis is placed on the strength, asset, or resource, and not on its nonuse.

4. *Weakness confrontation.* As the name implies, weakness confrontation dwells on the deficits of the person being confronted. Although it is impossible to avoid such confrontations entirely in the realistic interchange that takes place in a training group or in everyday life, weakness confrontations become the preferred mode of confrontation only of low-level communicators.

> *Group Member C:* Sheila, your silence here is beginning to drive me up the wall. It makes you seem so passive. When you're silent like that, I wonder what you're thinking; I begin to think that you're judging the rest of us. Your silence just doesn't get you anywhere. It makes you appear sullen and disinterested. And I want to block you out entirely.

Such confrontation may well reveal the impact that the silent member is having on many of the group members, but since it dwells merely on a weakness it probably does little good by itself. The person being confronted feels under attack and becomes defensive. My experience in groups tells me that the usual periodic weakness confrontation administered to a relatively silent member does practically nothing to increase such a member's participation. Nevertheless, group members continue to employ weakness confrontation in an almost ritualistic way.

> *Group Member D:* Sheila, you've mentioned that part of the problem you have with silence here is a fear of barging into conversations, interrupting others, and the like. Well, you can call it "barging in" if you like, but I *welcome* your interruptions. I guess I just don't see them as interruptions. I see them as your thinking enough of me to want to make contact with me. And I like that very much.

Here D tries to understand Sheila's feelings, but the focus is on a strength—Sheila's presence as an act of contact, an act of interest. Whenever you are about to challenge someone in the group, pause a moment and see whether you are emphasizing his strengths or his weaknesses. If a weakness is the primary focus, the odds are that your confrontation will be less effective than it could be.

5. *Encouragement to action.* As we take a look at this final type of confrontation, you will probably begin to notice that most confrontations are "mixed"; that is, they are combinations of elements from more than one type. Encouragement to action, as Berenson and Mitchell note, is my pressing you to act upon your world in some reasonable, appropriate manner and my discouraging in you a passive attitude toward life. High-level communicators are agents, doers, initiators; they are not afraid to make an impact on the lives of others. They are reasonably assertive and active, and they are unafraid to call others to action, especially when others say they want to act but fail to do so.

Group Member D: Ken, you and I have both admitted that we're not so assertive as we want to be in the group. I'm wondering whether you and I might enter into a little side contract with each other. Let's increase the frequency of our interactions. For instance, we can determine before a group meeting how often we would like to contact others, and then increase our contacts in a reasonable way. We could both help each other plan before the meetings and monitor each other's behavior during the meetings.

Not only does D confront Ken by encouraging him to act, but he suggests a way of going about it that emphasizes *mutuality.* Helping is going to take place, but it is *mutual* helping that doesn't cast either person in the role of "helper" or "helpee."

THE MANNER OF CONFRONTING

If confrontation is to be for better rather than for worse, the manner of confrontation is as important as the type of confrontation. First we will discuss some principles and then proceed to some practical applications of these principles. How should confrontation take place?

1. In the spirit of accurate empathy. We have already noted that even primary-level accurate empathy can be confrontational in itself. Advanced accurate empathy is in and of itself challenging and is therefore included among the challenging skills. However, the principle is that *all* interactions that take place among you and your fellow group members should be based on accurate understanding. If confrontation does not stem from understanding, it will almost inevitably be either ineffective or destructive.

Group Member E: If I can put together what you're saying, Mary, there seem to be two themes. You're attracted to George here and feel a good deal of respect for him, and you show this respect by making an effort to contact him, by working at understanding him, by giving him feedback on his style as he interacts with others. On the other hand, you haven't gotten over the fact that he ripped into you in a punitive, confronting way at the very first meeting we had. So you also still feel uneasy with him—there's still some distrust, alienation, and perhaps at times dislike. It seems to me that these negative elements choke off free-flowing communication between the two of you. I'm just wondering whether you both feel that it might be time to face these issues directly with each other. I think a number of us have taken on the role of intermediary between you two.

If E's assessment is accurate, it is likely to be constructively confrontational.

2. Tentatively. As is the case with all challenging skills, confrontation should take place tentatively, especially in the earlier stages of the lab experience, when you and your fellow group members are working at building initial rapport with one another.

> *Group Member A:* Jerry, you say that you swallow your anger a lot. Could it be that the anger you "swallow" here doesn't always "stay down"? From the feedback you're getting, it seems to dribble out somewhat in cynical remarks or periods of aloofness or uncooperative behavior. I'm wondering if you see some of this. I know you can deal with strong emotion directly. I've seen you do it here.

The fact that A's statement is filled with qualifications can help Jerry listen to and explore what is being said more easily. If you dump a ton of bricks on one of your fellow participants, you arouse his defensiveness, and he will have to pour his energy into recovering from the blow rather than into trying to assimilate and work with the confrontation. See how different the following confrontation sounds from the preceding one.

> *Group Member B:* Face up to it. You don't "swallow" your anger. It's dribbling out unproductively all the time. I don't think you're fooling anyone but yourself.

This is confrontation, but it is unnecessarily accusatory. Perhaps the word "balance" would be good to use here. Although good confrontations are not accusatory, neither are they so qualified and tentative as to lose their force. There is a message in the overly tentative confrontation: it says to the person being confronted "You probably don't have the resources to hear this." Such confrontations are demeaning because they turn the other group member into a "helpee." Likewise, the overly tentative confrontation may say something about the confronter: that he is afraid to face another person directly.

3. *With care.* Basic respect demands that group members confront one another with care. Let's try to operationalize the expression "with care."

Involvement. Confrontation should be a way of getting involved with the other person. If in the act of confronting you find yourself standing off from the other, you are probably not confronting with care. All interactions in a human-relations-training lab should heighten mutuality. If confrontation is alienating, it runs counter to the basic goals of the lab.

Motivation. Your motive in confronting should be mutual help in understanding interpersonal style and behavior. Other motives—to show that you are right, to punish, to get back at someone, or to put someone in his place—are dysfunctional. If your confrontation is not a way of showing others that you are "for" them, you will be seen as untrustworthy in the group.

The strength of the relationship. Confrontation should be proportioned to the relationship between yourself and the person you are confronting. We all know that we are more willing to hear strong words from some than from others. If you do little to establish rapport with your fellow participants, you should avoid the "strong medicine" interaction of confrontation. Caring confrontation presupposes some kind of intimacy between confronter and confronted.

The state of the person being confronted. Social intelligence demands that you be able to judge the *present* ability of your fellow group member to assimilate what you are saying. If the other person is disorganized and confused at the moment, it does little good to add to his disorganization by challenging him further.

> *Group Member A:* I'm just getting in touch with the fact that I practically never do give support here. I had no idea that I could be so selfish, but as I look at my style "outside," it's much the same. I get my own way. I take little or no time to understand others. I'm embarrassed. I'm almost speechless. It's like being groggy. I don't want to look at it for a while.
> *Group Member B:* I think you should see one more thing, while you're at it. Even though you don't give support, you demand it of us here. And that's infuriating!
> *Group Member C:* What has just dawned on you is pretty painful and confusing. It seems that you'd like a little time to let it all sink in, to settle down, to get your bearings.

C recognizes A's disorganization and tries to give him support by understanding what is going on (primary-level accurate empathy). B, on the other hand, moves in for the kill. His confrontation is not a sign of caring but a way of satisfying his own immediate need to express his pent-up emotionality and, possibly, to punish.

4. *Using successive approximation.* In many cases, confrontation will be more effective if it is gradual. The other person has to assimilate what is being said to him; he has to make it his own, or it won't last. If you are trying to change your own interpersonal behavior, it is usually not good to demand everything from yourself all at once. Space the demands you make on yourself; give yourself time to get a sense of success. This movement in small steps toward a behavioral goal, each of which is rewarded in some way, is called the method of "successive approximation." Therefore, if you are using encouragement-to-action confrontation, you will probably be more helpful if you start with simpler, concrete behavioral units that are relatively easy to change. Let's take a look at an example of what *not* to do.

> *Group Member A:* Jim, if you want to get rid of your feelings of loneliness, you have to get out there *today* and start interacting effectively with other people.

Member A is unrealistic: he is not concrete, and he asks for everything at once. Asking for everything at once is a good way of getting nothing. Let's look at another example, in which the group member is much more aware of the method of successive approximation. Bill is worried over the fact that he makes, or at least thinks he makes, a poor impression on others. The group has been meeting long enough for the members to have established a fairly solid degree of rapport with one another.

> *Group Member B:* Bill, you've said a few times here that you're pretty sure that you make a poor impression on people—that you seem, almost invariably, to start off on the wrong foot. First of all, it might be good to check that out with the people right here. I can give you some of my impressions. For instance, when we're together here, your posture frequently seems to say that you'd rather be someplace else. It could be that others would see you as being interested in them if you simply attended to them more carefully. At least that's my impression.

Although B may think that it is not just a case of poor attending that makes for Bill's poor impression on others, he starts there in his encouragement-to-action confrontation. Attending is a behavior that is relatively easy to change, and he challenges Bill to change his behavior here and now, in the group. Let's consider another example. In this case, a group member is too passive and ends up being ignored by the others.

> *Group Member C:* You're too passive, Ted. You have to go out and seize life if you expect others to pay any attention to you.

The concept "passive" is too general, and the solution offered is too vague. Moreover, C has assumed a "helper" or parental role; he no longer seems to be "with" Ted but is standing off from him.

> *Group Member D:* Ted, I have to admit that I ignore you here at times. When I ask myself why, I come up with a couple of things. I think enough has been said about your silence. But even when you do speak, your voice is so soft and quiet that sometimes it's hard to hear you. It's almost as if you don't want to touch me too strongly, even with your voice. I would feel better if you spoke up to me more.

D is involved and concrete, and he avoids the parental tones of C. He doesn't make wide-ranging demands but starts with something that may well be within Ted's control.

Some Practical Hints Concerning the Manner of Confrontation

If the effect of your confrontation is to set up a barrier between you and the other person, this confrontation is useless. Here are some ways to avoid erecting barriers.

Keep current. If another person's behavior is of some concern to you, don't keep your concern locked up inside yourself by bottling up your feelings, withholding feedback, ignoring the person, or engaging in other substitutes for directness. Try to stay current within the group; as issues come up, face them directly with the other person. This doesn't mean that you should become picky or specialize in challenging others. However, if you notice in another a behavioral theme that seems unproductive, don't hesitate to deal with it directly. For instance, if it becomes clear that another person is habitually passive or domineering in the group, don't wait until these patterns solidify and become part of the group culture. By that time it will be difficult to do much about them. Or, if you identify a resource or a strength that another uses only abortively, feel free to use a strength confrontation. In my experience, the majority of participants in human-relations-training groups (although there are disastrous exceptions) err by delaying confrontation too long—until it is useless or hostile—rather than by engaging in premature confrontation.

Nonverbal behavior is insufficient. Nonverbal hints about what you are thinking or feeling in the group are not enough. If you fall silent, refuse to look at someone else, exclude another, smile, slouch in boredom, or bite your lips, you are indeed sending messages—some of which may well be confrontational—but these messages must be backed up by words if they are to be direct and unambiguous.

Use describing behavior. Perhaps the most important practical hint for confronting others is one suggested by Wallen (1973): *describe* what you see as counterproductive behavior in another, and *describe* the impact you think it has on the person himself, yourself, and the other members of the group. There is a strong tendency, as Wallen notes, to substitute less useful forms of verbal behavior for descriptions. What are some of these?

Commanding Rather Than Describing

- *Commanding:* John, don't keep asking Mary how she feels. That's become a cliché with you. Tell her how *you* feel and what impact she has on you. Then she might do the same.
- *Describing:* John, unless I'm mistaken, almost every time you ask Mary how she feels—and you seem to do so often enough—I see her wince. You ask her to share her feelings with you, but I'm not sure I see you sharing yours with her.

Judging, Labeling, or Name-Calling Instead of Describing

- *Judging:* Peter, you're selfish!
- *Describing:* You're very verbal, Peter, and very assertive. At almost every group meeting you begin with your own agenda. We've almost developed a ritual here—we wait for you to start. I don't think that you or any of the rest of us is doing anything to change this pattern.

Accusing Instead of Describing

- *Accusing:* Gene, you have it in for Kay. You don't like her. You haven't accepted her from the start.
- *Describing:* Gene, I know that Kay has to work out her own relationship with you—but I see her sitting there and taking things that would drive me up the wall. You don't initiate conversations with her. When she contacts you, your part of the dialogue seems very brief. Even your voice strikes me as very matter-of-fact. I've noticed this kind of behavior from the very beginning of the group. I'm in the dark. I don't know what's going on between you two.

Questioning Instead of Describing

- *Questioning:* Jim, is it safe or productive for you to reveal so much about yourself so quickly?
- *Describing:* Jim, you have revealed yourself more than anyone else in this group. And now this evening you have begun to explore your sexuality—and it's only the third meeting! Frankly, it scares me when someone moves that fast.

Sarcasm or Cynicism Rather than Describing

- *Sarcasm:* Boy, you really know how to put people at ease!
- *Description:* Paul, now that Craig has stormed out of here, I'll bet you're wondering what angered him so. I don't want to talk about him, since he's not here, but this is my view of what you did. You told him point-blank that he's not the kind of person you usually make friends with, but you did it with such cool deliberation and lack of feeling in your voice. I shuddered a bit inside and began wondering what was happening inside Craig. Your expression hasn't changed since he's left, and I'm not sure what's happening inside you.

Describing behavior has a kind of objectivity to it that helps the person being confronted avoid the kind of natural defensiveness that arises with confrontation.

THE RIGHT TO CONFRONT

Does anyone and everyone in the training group have the right to confront fellow participants? It depends. By subscribing to the laboratory contract, you say in effect to your fellow group members "I want to place on myself the demands outlined in the contract. However, although I want to make these demands on myself, nevertheless I am a member of a learning community, and I expect my fellow group members to help me place these demands on myself without their becoming role 'helpers' or casting me in the role of 'helpee.' " Confrontation, then, is a legitimate form of interaction in the group, because it is one of the ways in which you and your fellow group members help one another live up to the legitimate demands of the training experience.

Does this mean that everyone in the group has an automatic right to confront? I think not. You must *earn* the right to confront others, and in general you can do so by living up to the contract yourself. Only active group members really earn the right to confront. Let's examine what this statement means more concretely. Berenson and Mitchell (1974) name certain qualities that must characterize a helper before he has the right to confront.

1. Relationship-building. In order to earn the right to confront, you must actively engage in the process of relationship-building as it has been described so far in this book. If you do not have a solid relationship with someone, don't confront him, for mere acquaintance does not supply a solid enough foundation for confrontation. Leave confrontation to someone who has a more solid relationship with the person. In order to confront, you must be "in" the group and not just a peripheral member.

2. Understanding at a deep level. Do not confront anyone until you have spent a good deal of time trying to understand him. Effective confrontation is built on understanding and flows from it. Without understanding, the confrontation is usually hollow. Moreover, unless the other person feels that you understand him, he will probably not listen to your confrontations anyway. Therefore, before confronting ask yourself whether you have spent sufficient time understanding the person.

3. Being able to disclose oneself. Do not confront others unless you are active in revealing yourself appropriately in the group. When you make yourself "visible" through self-disclosure, you too open yourself up to confrontation. The mutuality that should characterize transactions in the group demands that you disclose yourself. Don't expect to deal with the vulnerabilities of others without at the same time making yourself vulnerable.

4. Being in touch with oneself. The person who confronts should be in touch with his own emotionality, should have a feeling for the strength of the relationship between himself and the person to be confronted, and should know *why* he is confronting. If I am out of touch with my own experience, I may be confronting in order to "get even," in order to compete, in order to express my counterdependency, or for a variety of other counterproductive reasons. If your reasons for confronting are foggy to you, delay your confrontation unless you can work out your motivation with the help of the rest of the group.

5. Having a dynamic relationship with the confronted. Confront only those with whom your relationship is growing. If you and another person are on a kind of interpersonal plateau, delay confrontation and find out why nothing is happening between you. Unless you have a sense of the interpersonal movement of the other, your confrontation may well be stale or ill-timed. Confront only those with whom you have a currently active relationship.

6. Living fully. Berenson and Mitchell (1974) claim that only a person who is striving to live fully according to his value system has the right to confront another, for only such persons are potential sources of human nourishment for others. In other words, don't confront others unless you are the kind of person who challenges yourself.

7. Not allowing your virtues to be exploited. Don't allow your virtues to be turned against you in the process of confrontation. For example, if you are an understanding person, don't let the other be self-servingly selective in what he will listen to from you. If the other person is willing to be understood as fully as possible by you, he is also opening himself up to be confronted by you. Don't let yourself become the victim of the "MUM effect," which refers to people's tendency to withhold bad news from others (Rosen & Tesser, 1970, 1971; Tesser & Rosen, 1972; Tesser, Rosen, & Batchelor, 1972; Tesser, Rosen, & Tesser, 1971). In ancient times the bearer of bad news was often killed. In modern times he may not fear death, but he does fear something. Research has shown that, even when the bearer of bad news is assured that the one receiving the news will take it well, he is still as reluctant as the bearer who knows that the receiver will take it hard. Bad news—and, by extension, the "bad news" involved in any kind of confrontation—arouses negative feelings in the sender (the confronter), no matter what the reaction on the part of the receiver. If you are uncomfortable with human-relations training as a social-influence process, you may well fall victim to the MUM effect, and your communication with others will become watered-down and safe. Conversely, if you are the kind of person who earns the right to confront, you will probably engage in confrontation—not as a specialty, but naturally. Never confronting may be a sign that the quality of your other interactions is poor or weak.

8. Responding well to confrontation. Finally, confront only if you yourself have learned how to respond well to confrontations. What constitutes effective response to confrontation? Let's try to answer that question now.

RESPONDING GROWTHFULLY TO CONFRONTATION

Confrontation, even when it is executed with care by someone who has our interests at heart, has the tendency to pull us up short. It isn't always easy to be challenged. Therefore, we all have the tendency to respond defensively to confrontation. Since we are such resourceful creatures, this defensiveness can take many forms.

Defensive Response to Confrontation

Confrontation usually precipitates some degree of disorganization in the person being confronted. It can leave us feeling inadequate. One way of looking at confrontation and response to it is from the point of view of the cognitive-dissonance theory. Confrontation induces dissonance. For instance, I have always seen myself as a witty, humorous person, and now the suggestion is made that my humor is often biting or sarcastic, that it is a way of avoiding intimacy, and that it ignores the needs of others. All of a sudden I am forced to think of myself in a different way, and this shifting of gears is not easy. I am somewhat taken aback, pulled up short, and confused. This is dissonance. Since dissonance is an uncomfortable state, my immediate tendency is to try to get rid of it. But I can try to get rid of it in ways that don't allow me the opportunity of examining the confrontation in order to see if there is something to it. They simply protect me. What are some of these? (See Egan, 1975b; Harvey, Kelley, & Shapiro, 1957; King, 1975.)

1. Don't think about it. I can listen, say "I see," and then proceed to file what has been said in the deepest recesses of my mind. This seems to happen all too often in groups, and no one does much about it.

2. Distort the evaluation. Someone tells me that I am an attractive person and that my verbal skills are good, but that I don't use them assertively enough. I hear the part about my attractiveness and my verbal skills, but the rest is lost. In this case I distort the evaluation by not hearing the whole message.

3. Discredit the confronter. Another ploy to reduce dissonance is to destroy the credibility of the confronter. Counterattack becomes a strong defense.

> *Group Member A:* It's easy for you to ask me to become more involved here. You're attractive and skillful and can do easily what demands a great deal of effort from me. If you would stand in my shoes a little while, I don't think you'd be so quick in telling me what I'm doing wrong. Not only that, there are two people here whom you practically ignore. I think you should tend to your own garden first.

Obviously, you need not discredit the confronter so vocally. You can also do it silently, and just let him whistle in the wind.

4. Persuade the confronter to change his views. Win him over by persuasion or rationalization. Defuse the confrontation by linking up with the confronter.

> *Group Member B:* I know I don't express much feeling here, John, but I'm not sure that it's called for—not as much as you ask for, at any rate. After all, there *is* something artificial about these groups. I don't want to manufacture emotion; in fact, that's part of our contract. I want to let what comes flow naturally.

B's "reasonableness" here prevents him from facing up to the issue.

5. Devaluate the importance of the topic in question. This is another form of rationalization. For instance, if one of the group members is being confronted about his sarcasm, he may point out that he doesn't really mean it, that he is rarely sarcastic, that "poking fun at others" is a very minor part of his interpersonal style and not worth spending time on. The person being confronted has a right to devaluate a topic if it really isn't important. This fact emphasizes the necessity of your attending carefully to his behavior and understanding him before you confront him.

6. *Deny, reject the confrontation.* One defense is merely to deny what the confronter has to say. This approach, however, lacks subtlety, and more indirect defenses are usually preferred.

7. *Seek support elsewhere.* I can probably always find people in the group who are "on my side." Then, if I don't like what someone is saying to me, I can seek support for what I do from "my friends" or from those who find it difficult to say anything confrontational at all.

8. *Agree with the confronter.* I can readily agree with what the confronter has to say. But this too can be a game. It gets the confronter off my back, and my "honesty" wins approval from other group members. I find that, if I readily agree with the confrontation, that is the end of it. It is difficult for the confronter to confront me again, because I have already admitted what he has to say. The point is that I do little or nothing about it.

Obviously, there are many more defenses against confrontation, some blatant and some subtle. Use your imagination and list some further dodges.

Creative Response to Confrontation

If someone takes the risk of confronting you and does so reasonably and responsibly, then mutuality demands that your response be just as open and direct. What are the implications of this demand?

First, make sure you *understand* what the other person is saying. Use primary-level accurate empathy. Very often the confronter finds it difficult to say what he feels he should say. Respect for him demands that you get what he is saying straight, and that you understand how he feels as he says it. Obviously, your own emotions come into play. Since you don't like what you are hearing, it is easy for you to distort and misunderstand the message. You may have to fight your own emotions in order to understand. Confrontation, even when it is reasonable, can *feel* like an attack, and attack calls for instinctive defense and/or counterattack. Therefore, it may *sound* simple and easy to understand what the confronter has to say and make sure you get the message straight, but in practice it is not that easy.

Second, you are in a group, so use the resources of the group. If you are confused, or if you don't understand, deal with your confusion in the group. A good confronter often throws the question open to the entire group.

> *Group Member A:* Peter, you are a very active person in this group. Perhaps even overactive. Because you are so active, in both initiating conversations and responding when others talk about themselves, in some sense you end up controlling a lot of what goes on here. I wonder sometimes whether you possibly

avoid some of your own interpersonal issues by being so active. What I'm saying is a hunch. I'd like to check it out with the others here.

Peter: You're saying that anyone as active as I am has a kind of advantage. He can control his own agenda and thus, to a degree, he can control what is going to be said to him. You know, I may well be avoiding some interpersonal issues. I'd like to find out from the rest of you what some of these issues are—I mean the ones that pertain to this group.

Peter understands A and follows his suggestion by calling on the resources of the group. There is a big difference between calling on the other group members for one's defense and calling on them in order to explore the confrontation as carefully as possible. I have often been in groups in which both confronter and willing confrontee have tried unsuccessfully to get to the heart of the confrontation. But they didn't call on the resources of the wider group, and apparently no one else wanted to "interrupt." We will return to the question of how to muster the resources of the group in Chapter 12, when we deal with group-specific skills.

Third, respond to the confrontation, once you understand what is being said, by using the resources of the group in order to explore it as concretely as possible. For instance, if someone challenges you because you are a low-level initiator in the group, find out whether others see you in this way, find out how they feel about it, and explore how you yourself feel about it. Examine your own values related to being an initiator in interpersonal situations. You may find that you want to increase your level of initiation because your present level is so low that it interferes with the overall quality of your presence. However, you may also come to realize that you don't want to be an extremely assertive person. For instance, you may not want to be so assertive as the person who challenged you in the first place. Within the group you want to live up to the demands of the contract, but you also want to be yourself. In brief, whereas defensiveness is almost always counterproductive, mere compliance to the expressed or implied demands of the confronter is hardly the ideal. If you are to actively choose the elements of your own interpersonal style, you certainly should listen to the challenges presented to you by others; but you should also adapt these challenges to your own needs, wants, and values. If you determine that a given confrontation has merit, your ultimate response will be some kind of *change of behavior.* Often enough it takes time for a particular confrontation to sink in and take effect. Good response to confrontation doesn't mean leaping into action. Giving yourself time is useful, providing it is not just your way of sidestepping the issue.

CONFRONTATION AND MUTUALITY

Confrontation can be a very beneficial transaction when it is used reasonably and in a context of mutuality. Its obvious drawback is that the confronter becomes the "more knowing" person and the confrontee becomes the "less knowing" person, and mutuality is thus lost. Once a group is divided into "those who confront" and "those who are confronted," the group is no longer a learning community but a group of "helpers" and "helpees." Finally, as research has shown (see Berenson & Mitchell, 1974), anyone who makes confrontation his *specialty* does all things poorly, even his specialty.

Gordon (1970), in teaching parents ways of being effective in their relationships with their children, speaks of the "dirty dozen"—twelve categories of ineffective parental behaviors. Seen in the context of adult-adult relationships and human-relations training, these dozen sets of behaviors are caricatures or perversions of confrontation. As you read what follows, it may strike you that Gordon is overstating his case—that some of the behaviors he discusses are quite legitimate and do not destroy mutuality. Let's first go through the "dirty dozen" and then return to this issue.

- *Commanding, ordering, directing:* "John, talk to Jane. Your last remark hurt her." This set of behaviors constitutes an attempt to *control* the other. They suggest that the other person doesn't have enough social intelligence to be in control of his own interpersonal behavior. The implication is nonacceptance of the other.
- *Warning, admonishing, threatening:* "Jane, you know it's not good to pout like that. You'll just end up ostracizing yourself from the group." This remark certainly assumes that the other person is a child. These behaviors promote fear and submissiveness and usually evoke resentment and hostility.
- *Exhorting, moralizing, preaching:* "Peter, the contract is not an obstacle. It's here for our growth and protection. Following it actually frees us." Unfortunately, the temptation to preach in human-relations-training groups is sometimes quite strong. Sterile preaching effects little or nothing in religious contexts, so we should expect little from it in training groups.
- *Advising, giving suggestions, offering solutions:* "Sarah, why don't you relax and count slowly to five before responding to anyone who confronts you? I'll bet that will take the edge out of your voice." Again, this is helping behavior, with the inevitable "helper-helpee" role-casting.
- *Lecturing, giving logical arguments:* "Tony, first of all, you disclose little about yourself. Second, your attempts at accurate empathy are too widely separated. Therefore, you can't expect others to listen when you complain about their 'selfish' behavior." Teaching turns the other into a student, and, in most educational contexts, lecturing does not imply mutuality.
- *Judging, criticizing, disagreeing, blaming:* "Alice, it's your fault that Connie left in tears. You're too blunt and harsh. That behavior of yours is one of the reasons for the low level of trust in here." Judicial functions have no place in the training group. Perhaps more than other "parental" behaviors, they elicit feelings of inadequacy, inferiority, stupidity, and worthlessness. They provoke defensiveness, anger, and even hatred.
- *Approval behaviors such as praising, agreeing:* "Rich, you're the most understanding person in this group. You put yourself out for others, and that's great. More of us should do that." Such behavior usually makes the person being praised squirm in his chair and arouses resentment in others because of the expressed or implied comparisons. Praise can also be ingratiating and manipulative.

- *Name-calling, ridiculing, shaming:* "Dorothy, you really say stupid things at times. You act like such a little child." It goes without saying that such behavior seldom evokes beneficial change.
- *Interpreting, analyzing, diagnosing:* "Tom, from what you've said, I'd say that your fearfulness in groups stems from your childhood. You were too protected and sheltered by your parents. I think you still look for parent substitutes in this group." Playing psychologist is, again, a strong temptation. And it is endless, for when will we ever discover our "real" motives for doing things? The pursuit of insight too often leads nowhere, at least in terms of constructive behavioral change.
- *Reassuring, consoling, supporting, sympathizing:* "Cindy, don't let this get you down. I know we're making demands on you, but we care about you. We even respect your tears." Such behaviors make the other a "weak child" and usually cut off communication. They are distortions of accurate empathy and in no way encourage mutuality.
- *Probing, questioning, interrogating:* "Carl, how do you feel right now? Did Mark's comment embarrass you?" Questions usually beget little but more questions. Such cross-examination is demeaning. Questions imply that information is more important than understanding, that you are gathering data in order to "solve" a problem, and that *you* must solve the problem rather than let the other solve it. Questions limit the freedom of the other to talk about what *he* wants to talk about.
- *Withdrawing from, humoring, distracting:* "Kathy, let's put the issue of your relationship to Andy on the shelf for a while. Too many cooks here are spoiling your broth!" Humor too often implies a lack of interest and/or respect.

I don't mean to suggest that none of the behaviors above is ever useful in human interactions. I do suggest that parentalness is no substitute for mutuality. Some of these behaviors may be used beneficially once a certain degree of mutuality is established. The stronger a relationship the stronger the interactions can be—including strong confrontation. I include the "dirty dozen" here as a way of asking you to be careful in confronting. If you do use some of the "dirty dozen," it may help the interaction to consider your motivation for doing so. It may also help if you identify *what* you are doing, so that the person being confronted does not feel attacked. Finally, any kind of confrontation is suspect when it is used as a substitute for understanding.

SELF-CONFRONTATION

Challenging one another in the group has been emphasized up to this point. However, mutuality is well served if you and your fellow group members learn how to challenge yourselves. Put crudely, you should "get yourself before others get you." This statement is crude, because responsible confrontation is not a "getting." But there is a great deal of value in this idea. If I learn how to confront myself, I will be less defensive in dealing with the issues I bring up, will not force others into the position of constantly challenging me, and will put the responsibility for change directly where it belongs—on myself. If I challenge myself, I can then call upon the resources of the other members of the group to help me monitor my behavior, to give me feedback, and to provide some of the encouragement and support I need to work through a change-of-behavior program. Ideally, then, self-confrontation should be more frequent than confrontation by others (or at least should grow in frequency). For most of us this means that we must adopt new behavior. The laboratory is precisely what we need—a place where we can experiment with new behavior under controlled and supportive conditions.

EXERCISES IN CONFRONTATION

This section will describe exercises that will help you get some kind of experiential feeling for confrontation, good and bad. Exercises in confrontation that are more directly related to the group experience itself will follow, when human-relations training is dealt with specifically as a group experience.

Exercise 42: Self-Confrontation, Good and Bad

Before confronting others, it is best to practice on yourself, for misguided and irresponsible confrontation can do a great deal of harm. The purpose of this exercise is to give you a feeling for both responsible and irresponsible confrontation.

Directions

Think of a few areas in your interpersonal life in which you could benefit from some kind of challenge or confrontation, areas in which you should be invited to examine your behavior more carefully. Then write out a statement in which you confront yourself irresponsibly. Next, write out a statement in which you confront yourself responsibly—that is, tentatively, as an invitation to self-examination. First, study the following example.

Example

a. "Why do you insist on being so parental? You recognize that you are controlling, and you seem to regret it, but you do absolutely nothing about it. It's damned arrogant and insulting. I don't think that you have a basic respect for people. Sure, you're an attractive, flashy guy, but in the core of you there is a great deal of selfishness and inconsiderateness. Your need to control and get your own way is part of the structure of your personality and not just a set of learned behaviors" (irresponsible).
b. "I know that others have talked to you about being parental and controlling, but, since it bothers me, I think it's only honest for me to bring it up. I guess I

think that it has been brought up but not looked into very deeply. I see you as a very strong and very talented person. It's almost as if you use your strengths against yourself—for, when you try too hard to control, I believe you alienate. At least I tend to withdraw. Words such as 'selfish' and 'inconsiderate' have come up—and I really don't know whether they apply to your style or not. I do wish we could take a longer look at some of the issues that bother me and perhaps others here" (responsible).

Write out four pairs of self-confrontational statements like those above, one irresponsible and one responsible. Choose areas related to your interpersonal style, areas you would like to work on.

In a round robin, share your self-confrontational statements with four different members of your group. Check to see whether your "responsible" confrontation is seen as such by your partner. What words or phrases introduce tentativeness? Is the confrontation concrete enough to lead somewhere? Point out the elements that make the "irresponsible" confrontation irresponsible.

Exercise 43: Changes from First Impressions

One way of giving feedback to your fellow group members is to give each one "current impressions"—that is, concrete impressions of how they strike you at this time in your relationship. Sometimes members of human-relations-training groups exchange "first impressions." There are advantages and disadvantages to such a task. One disadvantage is that first impressions are often superficial and transitory. However, a possible advantage is that people are, *de facto,* affected by their first impressions. Some people seek a person out again, and some avoid him, on the strength of their first impression. It may be good to find out the nature of our first impressions upon others, both positive and negative. Such an exchange of impressions tends to be confrontational. This is all right if the general rules of good confrontation are observed (caring, involvement, concreteness, tentativeness, description of behavior, and so on). Perhaps more important than first impressions, however, is the way in which these impressions change as relationships grow. Therefore, this exercise will deal with that process of change, combining first impressions and current impressions.

Directions

Think of the first impressions you had of each other member of your group. Write these down. Then think of how these impressions either were confirmed or were changed as you got to know the other person better. For each person, write something positive and something negative. However, choose something you think will help the other person pursue the goals of the group. Study the following examples.

Example 1

John, my first impression of you was that you were a shy, rather passive, but "nice" person. You seemed to hesitate to speak up, to initiate, to take responsibility for the direction of the training process. I noticed early in the lab that you blushed once or twice when others called on you to be more assertive. This confirmed for me my impression of passivity or shyness. However, as the group moved on, what I thought was weakness proved, for me, to be strength. You don't interact hastily, but I see in your deliberateness a deep respect for others. When you do interact, it's with intensity, conviction, and care. I sit up and listen when you talk to me. You're not "showy," but you're not passive. I've seen you get angry, control your anger, and then actually use it to make some transactions more meaningful.

Example 2

Martha, my first impression was that you were very attractive physically, and because of your physical attractiveness I think I ascribed to you all sorts of moral qualities. That was probably unfair of me. I saw you as vitally interested in everyone. Active. Socially intelligent. Obviously, that was a lot to live up to. Since then I've become more realistic. I guess I don't see you as quite that altruistic. You've had a hard time getting to know Jill and Sam here, and at times I'm not sure that you've expended enough energy to do so. I've seen you confront Bob poorly—or at least it seemed to me—on a couple of occasions, so, although I see you as socially intelligent, I also see you as fallible. You do make mistakes, and you're just as prone to laziness as the rest of us. I see you invested in your own growth, though. You do usually come in here with an agenda; there are issues with yourself and with others that you have thought about and want to do something about. That increases my respect for you.

Do one of these comments in writing for each of your fellow group members. You need not give a rundown on the other person's entire personality. Choose one area or one item that you think may have significance for the other or that has significance for you. Use the formula

In the beginning I saw you as . . . , but now I see you as . . . , and these specific things have caused my impression to change.

Or:

In the beginning I saw you as . . . , and this impression has been confirmed for me by the following experiences.

This is not feedback for the sake of feedback. Try to make your statements useful by helping the other explore his interpersonal style and by providing background for the present state of your relationship with him.

You can share your statements with the others either by means of a round robin or openly with the entire group. Obviously, the second process will take much longer.

Try to listen to the feedback carefully. Communicate understanding to the person giving you feedback. Clarify the feedback if it is not clear. As you get feedback from each individual, listen for recurring themes (such as unused ability to be assertive, a tendency to surrender the directionality of the group to the leader or to others, inappropriate warmth that meets your needs rather than the needs of those to whom you are being warm, and so forth).

Exercise 44: Self-Confrontation:
The Quality of My Participation in the Group

Since you should make demands on yourself, and not merely wait until others place demands on you for effective group participation, you should learn how to examine and monitor the quality of your group participation. This exercise asks you to take a look at your principal underdeveloped strengths and the weaknesses, discrepancies, and distortions that may be part of your style.

Directions

Review your participation in the training group. See if you can find one or two things to confront yourself on in each of the Berenson and Mitchell (1974) categories. Remember, these categories are:

1. Didactic confrontation
2. Experiential confrontation
3. Strength confrontation
4. Weakness confrontation
5. Encouragement to action

Study the following example.

Example

- *Didactic:* I don't understand the function of the leader. I am especially confused about whether he should be putting himself on the line in the same way as the rest of us. Up to this point I haven't cleared this up, and therefore I hesitate to initiate anything with him.
- *Experiential:* There are discrepancies. For instance, sometimes when I'm bored I assume an attending position in the group, in truth not to get involved but so that I won't be called down for lack of involvement.
- *Strength:* I have the ability to initiate with relatively nonparticipating members in ways that don't alienate them. We have two underparticipating members in our group. However, I have done little up to this point to set them at ease and share myself and my own misgivings. I see potential in them, but I don't use my initiating resources to get at it. I'm lazy—or have been up until now.
- *Weakness:* When anyone challenges me I disarm them by smiling, being warm, and giving the appearance of openness. I use challenges as an opportunity to look only at those areas *I* want to look at.

- *Encouragement to action:* I want to "blow my cover" and reveal the "challenge-response" game I've been playing. I think this will be somewhat embarrassing, for I've played it a number of times and duped others by my façade of openness. Perhaps the best time to reveal it is the next time I'm challenged. I can show others what my instincts tell me to do and then try to be authentically open.

This exercise should give you plenty of material for self-confrontation in the group. You can use this exercise repeatedly to provide yourself with self-confrontational material. You can share it in the total group systematically or merely use it as grist for your mill in your ongoing interactions.

Exercise 45: The Confrontation of Strengths

As I have already mentioned, high-level communicators confront the strengths of others much more frequently than their weaknesses. This exercise is designed to direct your energies into doing precisely that.

Directions

In your mind's eye, successively visualize each member of your group working "at his best" in the group. What do you see for each person? Obviously this visualization is an expression of your hopes for each, but presumably your hopes are based on the potential of each person in the group. Write down a few of the things you "see" each person doing well. Be as concrete as possible. Study the following examples.

Example 1

I see Paula risking herself in terms of self-disclosure. I know that right now she's hesitant and cautious, but I think I see a kind of basic courage in her. I see her taking a special risk in the disclosure of her thoughts and feelings on what is happening in the here and now of the group. Her relationships with the other members of the group grow stronger each week. She now initiates with practically everyone. I see this initiation as the basis of her ability to be more self-revealing. She is building a secure base.

Example 2

Even though John started out on the wrong foot with both Rita and Ken, I see him as capable of 'mending fences' and developing relationships of some substance with these two people. He alienated them by challenging them without having first communicated understanding. Even though they more or less rejected him by not initiating communication with him, he has remained open. He joins in conversations they have with others, and recently he has initiated conversations directly with both of them. I see him as capable of facing the issue of alienation directly in the very near future. A certain good-hearted simplicity got him in trouble with them, yet I think this same simplicity will be a resource to help him work his way out.

Again, it is not necessary to spell out every partially used resource the other person has. As you review in your mind's eye each person "at his best," choose one or two talents, abilities, and/or resources that you believe would be important for the person to develop.

Share your strength confrontation with each member of the group. This can be done in a round robin or in the total group. As you yourself receive feedback, see if any themes of underdeveloped talents/abilities/resources emerge in your style. Do you see yourself as having the capabilities others see you as having? Do you want to develop these resources?

Exercise 46: Confrontation: Safe and Tension-Producing Areas

Not all interpersonal issues carry the same weight. For instance, each of us has relatively sensitive areas, areas we find hard to talk about. Your being in touch with both your own and your fellow participants' areas of tension will help you in the process of challenging others and opening yourself up to challenge.

Directions

1. Write down two lists for yourself: one, "safe" interpersonal topics; the other, tension-producing interpersonal topics.
2. Write these two lists also for each of your fellow group members—areas you see as "safe" for each and areas you see as tension-producing for each.

Example

- *Safe interpersonal topics for me*

  my self-image, how I feel about myself
  my relationships to everyone in the group except Bill and Barb
  how I am developing my interpersonal skills
  increasing my assertiveness both inside and outside the group

- *Tension-producing areas for me*

  sexuality
  my tendency to become dependent on others
  my relationships to Bill and Barb
  my defensiveness when any of the three areas above comes up

- *Safe interpersonal topics for Tom*

  his interpersonal values, at least in the abstract
  his present relationship to almost anyone in the group
  the way he is developing interpersonal skills
  the way others see him—that is, generally as attractive, warm, interesting, caring

- *Tension-producing areas for Tom*

  his self-image

his passivity, his low interest level
his investment in interpersonal change when it comes to concrete issues, such as taking greater initiative with others
his defensiveness, the way he backs down from interpersonal challenge

This exercise is meant to help stimulate your thinking about yourself and others. You can turn it into a formal group exercise by processing it in a round robin.

CHAPTER 10: FURTHER READINGS

Berenson, B. G., & Mitchell, K. M. *Confrontation: For better or worse.* Amherst, Mass.: Human Resources Development Press, 1974.

Carkhuff, R. R., & Berenson, B. G. In search of an honest experience: Confrontation in counseling and life. Chapter 11 in *Beyond counseling and therapy.* New York: Holt, Rinehart and Winston, 1967. Pp. 170–179. Chapter 11 is a collaborative effort under the direction of John Douds and with assistance from Richard Pierce.

Egan, G. Confrontation in laboratory training. Chapter 9 in *Encounter: Group processes for interpersonal growth.* Monterey, Calif.: Brooks/Cole, 1970. Pp. 287–335.

Chapter 11

Immediacy: Direct, Mutual Talk

Immediacy is one of the most important, but also one of the most difficult, of the interpersonal skills. There are two related kinds of immediacy: relationship immediacy and here-and-now immediacy.

RELATIONSHIP IMMEDIACY

This type refers to the ability to discuss with another person directly where you stand in your relationship to him or her and where you see the other standing in his or her relationship to you. In this context, relationship is considered a dynamic process rather than some kind of static entity. Since one of the operational goals of the lab group is for the members to establish and develop relationships with one another, the skill of relationship immediacy is critical.

> *Group Member A:* Jack, you and I haven't talked directly to each other for two or three sessions now. I notice that we—or at least I—tend to avoid even nonverbal communication. It's almost as if we had a pact: "I won't notice you if you won't notice me." I feel we are on a plateau. We revealed quite a bit about ourselves early in the lab by talking directly to each other. Then I'm not sure whether fear set in or whether we thought we were getting too committed to each other—I haven't even tried to explore it myself outside the group. But now I'm beginning to get uncomfortable. I don't know what you're thinking anymore, and that's bothering me. I'm hoping that our talking directly to each other early in the group wasn't phony, some sort of first fervor.

Here the emphasis is on the whole sweep of the relationship and on where the relationship itself stands right now.

HERE-AND-NOW IMMEDIACY

This type refers to the ability to discuss with another person what is happening between the two of you in the here and now of an interpersonal transaction. Let's look at an example.

> *Group Member B:* Lisa, I hear a cynical edge in your voice, and somehow I think it's aimed at me. I'm not even sure that "cynical" is the right word, but the tone is negative. Tonight I've been waiting for you to contact me directly, and I guess I'm still waiting. For all I know I have a lopsided view of what's going on between you and me, but I won't know until we talk. I'm angry right now, but I don't want to withdraw.

In these remarks, B is not dealing with the whole sweep of a relationship. Rather, he is concerned with what is or is not happening between himself and Lisa right now. Obviously, the sum of such here-and-now transactions ultimately defines to a great extent the quality of the overall relationship, but the two can be separated and indeed often should be. Some people make unwarranted generalizations about the quality of an interpersonal relationship from a single negative here-and-now transaction. For instance, a young couple might be having a disagreement that involves in-laws. She says angrily:

> You've never respected my mother. You don't respect her now, and you don't really respect me. I'm all right as long as I stay in my place and keep my family in its place.

Although her anger may give her an opportunity to express doubts and frustrations that have been overlooked or concealed up to the present, that anger may also lead to an overstatement of her case. Often enough we use words to punish one another, and this may be the case here.

IMMEDIACY: A COMPLEX SKILL

One reason immediacy is not easy is that it is a complex skill that demands a fair degree of social intelligence. Like other human-relations skills, it has three requirements: (1) perception or discrimination, (2) technological skill, and (3) courage.

The Perception or Discrimination Component

If you are going to talk to someone about what is happening between the two of you—either in your overall relationship or in a here-and-now transaction—you have to *know* what is happening. For instance, if you don't notice the nonverbal or paralinguistic cues from another person who is angry with you, you can hardly address the issue of the other's anger directly. "I had no idea she felt that way" is sometimes used as a defense, but it could just as well be a condemnation. Someone might well say "The reason you didn't know she felt that way is that you failed to attend and to listen. She gave a lot of cues. You just didn't pick them up." Immediacy, like other human-relations skills, demands good perception or discrimination.

The Technology Component

Immediacy is a *communication* skill formed from a combination of three other skills. For this reason it is called a "complex" skill. Immediacy is a mixture of accurate empathy, self-disclosure, and confrontation.

Accurate empathy. You must not only perceive what is happening between yourself and the other person, but you must be able to put your perceptions or understandings into words. Very often immediacy calls for *advanced* accurate empathy, for what is happening in the relationship is often not expressed openly and directly.

> *Group Member C:* Unless I'm mistaken, your tone with me the past couple of days has been matter-of-fact. Also, you're quiet, and sometimes you're quiet when something is bothering you. I'm putting these two together and asking myself whether there's something wrong between you and me.

Advanced accurate empathy does not require that you become a detective, piecing together the evidence. It does require that you notice behavior, including feelings, and try to communicate an understanding of that behavior. The accuracy of your perceptions arises not from trying to play the "I'm right, you're wrong" game but from the fact that you are invested enough in the relationship to be sensitive to all cues and messages.

Self-disclosure. Being immediate means revealing how you think and feel about what is happening in your relationship with the other person. Because you are "listening," in the widest sense of the term, you "hear" many messages and react to them. Through immediacy, you disclose these reactions to the other person. Let's continue the example begun with Group Member C above.

> Anyway, I'm beginning to get uneasy. I'm asking myself if I have given any signs of taking you for granted. I *feel* close to you, but I'm at a loss as to how to express it. I get to feeling awkward and then I do nothing. This must give the impression of taking you for granted—or at least that's my guess.

Immediacy is *not* a way of "dealing with" the other person. Rather, it is an exercise in mutuality; it is the beginning, hopefully, of dialogue; and it is an expression of the give and take of a relationship. The self-disclosure element of immediacy has a message: "I want to be my 'transparent self' in my relationship to you." Such disclosure, of course, must be appropriate if it is to be responsible and if it is to promote rather than retard the relationship. For instance, within your training group this means that you should be open in your relationships, but not that you necessarily should reveal the same kind of material you would reveal to an intimate friend outside the group. The criteria governing appropriateness of self-disclosure, discussed in Chapter 2, are equally valid for the self-disclosure that is part of immediacy.

Confrontation. Immediacy is always an invitation to enter into dialogue about the relationship, and therefore it has a challenging or confrontational quality about it. Immediacy is an act of intimacy, and it demands intimacy from the other. This demand can certainly be challenging. Let's return to C's statement of immediacy.

> I think that maybe we would both like to clear the air. I'd like to know whether any of this is making sense to you, and I'd like your help in getting my own feelings straightened out.

The invitation to explore a relationship mutually is perhaps always understood in any kind of statement of immediacy, but at times it is good to state the invitation directly.

It follows that if I am poor at any of the skills named above I will be poor at immediacy. But, since immediacy is central to relationship-building, both in training groups and in day-to-day living, I can't really afford to be poor at immediacy if I want to grow in my relationships.

The Courage Component

Immediacy is not just a question of good perception coupled with good communication technology. Perhaps more than any of the other interpersonal-communication skills, immediacy demands "guts"—courage or assertiveness. Too many people wait too long; that is, they allow a relationship to deteriorate in a relative or absolute way and then find that they don't possess the kind of courage that immediacy demands at that point in the relationship. For instance, various annoyances crop up between a husband and a wife, and they vaguely notice an uneasiness in the air. Nothing is said, however. Communication grows stale or edgy, but still nothing is said directly about the relationship. Now minor annoyances begin to become major, motives are questioned, insignificant behaviors are assigned sinister meanings, and feelings keep building up—until all of this breaks out in a game of "uproar" (see Berne, 1964). They fight, and emotions are spilled out in hurtful, irresponsible ways. Finally, their anger is (we hope) spent, and they settle down into their usual relationship pattern or at least enter into an uneasy truce.

The courage that is necessary to face such an interpersonal rift in the beginning, when it is small and manageable, seems to be worth the effort, unless one or both find some kind of unhealthy satisfaction in successive games of "uproar." Games keep people from real intimacy. Therefore, if this couple wants to avoid intimacy and its demands, they may prefer a games approach to interpersonal living.

IMMEDIACY AND THE "INTERPERSONAL GAP"

Even though we don't reflect on the fact very often, communication is a complicated process, and many things can go wrong with it. It contains the following elements:

- *Intention:* when you address someone else, you have some goal in mind.
- *Encoding:* you have to translate your intention into some kind of language—written or oral, verbal or nonverbal.
- *Decoding:* the receiver sees the message from his perspective and decodes it, assigning an intention to it.
- *Effect:* the totality of what has happened has an impact on the receiver.

Wallen (1973) calls the space between your intention and the impact your message has on the receiver the "interpersonal gap" (see pp. 218–224), and he discusses methods of closing this gap in growthful ways. Immediacy plays an important part in bridging this gap. But first let's take a closer look at the gap itself and give an example of how communication, even well-intentioned communication, can go awry.

The process	*An example*
1. *The speaker's private intentions.* You, in speaking to someone else, have certain private intentions (that is, wishes, wants, hopes, desires, or fears) that give rise to your actions. In terms we have seen before, these are *covert* behaviors or experiences. They remain private unless you reveal them.	You are listening to a fellow group member. You *want* to encourage and support him, "be with" him. You see that he is nervous as he talks about himself and the difficulties he has in relating to others. You *want* to experience "with" him, understand his world from his perspective.
2. *Transforming or encoding.* Next, you encode your intention, whatever it may be, in some sort of verbal or nonverbal behavior (or combination). You expect your behavior, verbal or otherwise, to be the vehicle of your intention.	You communicate, as best you can, accurate empathic understanding of what he is experiencing. This is your verbal way of "being with" him, of giving him encouragement and support.
3. *Transforming as decoding.* The other person experiences your verbal and/or nonverbal behavior. This means that	The person to whom you are speaking sees your accurate empathy as confrontational (placing a demand on him)

he may assign an intention to it *different* from your original intention.

4. *Private effects on the listener.* The impact the message has on the receiver will depend on how he decodes the speaker's behavior. If he decodes it as the speaker means it to be decoded, the private effects will be the same as the speaker intended.

and as an act of condescension (you are "helper" and he is "helpee").

The person to whom you are speaking feels put down and inferior. He feels you are pushing him. He is annoyed and resentful. He wishes you would keep your remarks to yourself. These private effects are covert experiences and behaviors.

If you are to develop the skill of immediacy, you must be aware of the complexity of the communication process and be in touch with your own communication behaviors. In this instance, immediacy demands being in touch with your own typical "encoding" and "decoding" behavior.

> You may be unaware of the ways you code your intentions and decide others' actions. One of the important objectives of this study of interpersonal relations is to help you become aware of the silent assumptions that influence how you encode and decode.
>
> If you are aware of your encoding operation, you can accurately describe how you typically act when you feel angry, affectionate, threatened, uneasy, etc.
>
> If you are aware of your method of decoding behavior of others you can describe accurately the kinds of distortions or misreadings of others you typically make. Some people, for example, respond to gestures of affection as if they were attempts to limit their autonomy. Some respond to offers of help as if they were being put down. Some misread enthusiasm as anger.
>
> Because different people use different codes, actions have no unique and constant meaning, but are interchangeable [Wallen, 1973, p. 221].

The communication process becomes complicated, and the possibility of misunderstood communication increases, for several reasons that Wallen discusses:

1. *The same intention may be expressed by different actions.* For example, you can let one of your fellow group members know you care about him by attending, by communicating AE I, by revealing yourself to him, by helping him make reasonable demands on himself to fulfill the lab contract, by reinforcing him when he takes a reasonable risk, and so forth. However, it is also possible that any of these actions may *not* be seen as acts of caring.

2. *Different intentions may be expressed by the same action.* You can use the communication of accurate empathy to

- show care and respect;
- show off your newly acquired skills;

- get a person on your side;
- make yourself a "helper" and the other person a "helpee";
- make yourself liked; or
- avoid reasonable confrontation.

3. *The same action may lead to different effects.* For example, if you confront one of your fellow group members, he might

- be thankful that someone is talking directly and seriously to him;
- see you as judgmental and punitive;
- see you as putting on a good show for the group;
- enjoy it, seeing that it's your way of telling him you like him; or
- tell himself that he might as well stop trying.

4. *Different actions may have the same effect.* For example, you might act in any of the following ways toward one of your fellow participants:

- tell him that you appreciate his attempts to make contact with you;
- point out ways in which he is failing to live up to the contract;
- take his side in an argument in the group; or
- blow up at him and chew him out for his passivity.

In *all* of these cases, the person being communicated to might feel pleased, saying to himself "He's paying attention to me; he must like me."

As Wallen notes, "Bridging the interpersonal gap requires that each person understand how the *other* sees the interaction" (Wallen, 1973, p. 222). In other words, accurate empathy (both AE I and AE II) is essential to immediacy.

IMMEDIACY AND THE INTERPERSONAL LOOP

If I speak to you and you then respond to me, we have made an interpersonal "loop." Let's take a look at an example of how communication can go awry along this loop.

I speak:

- *Intention:* I want to let you know that I care about you.
- *Encoding:* I choose to let you know this indirectly. I do so by revealing to you in the group something I feel strongly about but have kept to myself, both within the group and outside. I tell you how trapped I feel at work. I don't like the work, and my boss doesn't try to understand me at all. I feel alone and helpless. It is difficult for me to reveal these feelings. I take my self-disclosure to you as a sign of

intimacy. To me it says "I trust you. I care about you because I see you as trustworthy."

- *Decoding:* You see my disclosure as something "then-and-there" and irrelevant to what is happening in the group. You notice that I make no attempt to relate it to this group of people in this situation. You see what I am doing as inappropriate and as a relatively harmless but time-consuming deviation from the contract.
- *Internal impact:* You are annoyed, and you feel that I am running away from the immediate business of the group. You feel unprepared to respond to the content of my disclosure.

You respond:

- *Intention:* You want to let me know that my disclosure doesn't fit into what is happening in the group, no matter how meaningful it may be in itself. You feel that I should relate it to this group of people in this situation or drop it. You don't want to hurt me, though.
- *Encoding:* You confront me as carefully as you can: "I'm not quite sure how what you've just said about yourself fits into the group, Fred. I'm scratching my head and trying to relate it to us in some way, and I'm not getting very far."
- *Decoding:* I feel that you have ignored the real message: I trust you. I see your confrontation as putting me off. I see you as acting impersonal and adhering too rigidly to the contract.
- *Internal impact:* I feel hurt and alienated.
- *External reaction:* I mutter something about its not being that important and fall silent. I leave the group feeling empty and disappointed.

How could immediacy-type statements intervene in this process and make it more productive? In a number of ways. "I" could have done things differently, and "you" could have done things differently, too.

My encoding: I could have been more immediate.

> I feel awkward. I want to let you know that I care about you, but I'm at a loss as to how to do it except through the direct, blunt approach I'm taking right now. I don't want to offend you by my bluntness.

The encoding itself can carry the "me-you" message directly.

Your decoding: Seeing that my self-disclosure didn't fit into what was happening in the group, you could ask yourself "What is he trying to say to me? He's opening himself up to me, even though the disclosure itself is inappropriate."

Your internal impact: You might have two sets of feelings: (1) annoyance at the fact that my disclosure sends the group off on a tangent, but also (2) puzzlement about what I am trying to say to you.

Your intention: You choose to find out what is happening between you and me. You don't want to confront until you know what is happening.

Your encoding: This can be an immediacy statement rather than a confrontation.

> Fred, my first reaction was to ask you how what you've just said fits into what's going on in the group. But right now I guess I'm struck by the fact that you can trust me enough to let me know what you've been keeping locked up inside. I'm not exactly sure how to respond, even though I appreciate your confidence.

You receive the message I was trying to send, but you also let me know that it's hard to handle because it is so indirect and out of context.

Immediacy deals with

- my behavior and the impact I see it has on you
- your behavior and the impact it has on me
- your behavior and the impact you see it has on me
- my behavior and the impact it has on you
- our goals in relating to each other

Like confrontation, immediacy is best expressed through the description of behavior and the description of the impact this behavior has.

- "I see you frowning; I think my last remark probably sounded cutting to you."
- "This is the second time you've cut me off. I'm beginning to feel annoyed."

With the high-level communicator, immediacy takes the place of

- *name-calling*

  not: "You're a beast!"
  but: "When you confront me, you don't qualify what you say at all. It's not tentative, and I resent that."

- *expressions of approval*

  not: "You're wonderful!"
  but: "If you're not sure what I'm saying or feeling, you check it out with me. I like that. I feel I trust you."

- *accusations*

  not: "You never listen to me or anyone!"
  but: "You missed what I just said. And it hurts."

Cautions. Immediacy does not mean that you are forever second-guessing others, looking for hidden interpersonal meanings, and constantly placing your relationships under a kind of microscope. Immediacy, like the other skills, becomes a weapon (or merely a bore) in the hands of the inept. Low-level communicators are owned by interpersonal skills. They become technologists, executing the technology of the skill well, but losing the person. High-level communicators own the skills and use them to make more effective contact with others. The laboratory, as I have noted, does possess a degree of artificiality. We cannot escape the fact that the lab requires a focus on skills, relationship-building, and examination of relationships that is missing in day-to-day life. In a sense, the laboratory exaggerates (necessarily) certain dimensions of life that are too easily ignored outside the laboratory.

ONGOING OR INTEGRATED IMMEDIACY

When we focus on immediacy in itself, as we do in this chapter and in the exercises that follow, immediacy is taken out of context. It is seen as something added to interpersonal transactions. Ideally, however, the high-level communicator integrates immediacy into his transactions in an ongoing way. Relationship immediacy ("Where do we stand currently in our relationship to each other?") does not become a "heavy" issue (like a tune-up, or even an overhaul) that he must face after a certain amount of time has elapsed, for he feels free to engage in here-and-now immediacy ("What is going on right now as we relate to each other?") as it is called for. The high-level communicator *works* at his relationships, without becoming preoccupied by them.

EXERCISES IN IMMEDIACY

The exercises below are designed to help you focus on the relationship-building process with your fellow group members. They assume a relatively high degree of motivation to become involved with others. They are not magical substitutes for the work that must go into building relationships.

Exercise 47: Similarities and Differences

This exercise is designed to help you sharpen your perception as it is related to immediacy. Similarities and differences in personality traits and behavior affect the entire relationship-building process. Can you allow others to be different from you? Do you immediately try to make friends with someone who is similar to you?

Directions

Think of each member of your group in succession. Write down a few significant ways in which you think you are like each of the others in the group, and a few significant ways in which you think you differ from each. Be as concrete as possible. Study the following example.

Example

- *Ways in which I am like John:*

We are both rather quiet in the group, although not withdrawn.
We are both soft-spoken.
We both wait too long before we enter conversations taking place within the group, and we both wait until others contact us instead of reaching out and beginning conversations.

- *Ways in which John and I differ:*

When he does speak, he emphasizes accurate empathy, whereas I tend to talk about myself.
I have a quick temper, which I generally keep under control.
I fear the trainer and express it by being resistant to exercises, whereas John is also fearful but manifests his fear through compliance.

Next, briefly write down how you think these similarities and differences affect your relationship (or the relationship-building process) with each of the members of your group. For example:

Because of our similarities, John and I haven't gotten to know each other very well. We don't make contact easily, and we never challenge each other.

Because of our differences, I think John has a rather negative view of me. My guess is that he doesn't like my self-centeredness, my temper, or my resistance. I guess I think he is too "nice." He makes me feel guilty.

Our interaction has not been strong enough for any of this to surface very often.

In a round robin, share your reflections on each person with that person. What differences do you find between his views and yours? What are the implications for your relationship? What would you like to change?

Exercise 48: "Interpersonal-Process Recall": An Exercise in Immediacy

If you are to integrate immediacy into your interpersonal transactions in the ongoing way suggested in this chapter, you must learn how to become aware of immediacy issues as they come up. This exercise is based on a training technique, developed by Norman Kagan (1971), called "interpersonal-process recall." It is designed to help you become aware of immediacy issues as they arise.

Directions

Have a ten-minute conversation with one of the members of your group. Let another group member act as as observer. The topic of the conversation should be related to the goals of the group (for instance, interpersonal style, how I feel about my use of interpersonal skills, and so on). At the end of ten minutes the conversation is stopped, and the observer conducts an "interpersonal-process recall" interview. That is, he helps both of you explore what you were thinking and feeling about what was taking place in your conversation but *did not verbalize* to each other. These unverbalized dimensions of the conversation should be recalled *as concretely as possible* (specific feelings, experiences, behaviors).

Study the following example. Bill and Steve talk to each other for ten minutes; at the end, Carol, the observer, helps them verbalize what didn't come up directly in the conversation.

> *Steve:* When I touched on my sexuality, I felt I was getting in too deep. I wanted to pull back, but I felt awkward doing so—as if you knew that I was trying to get out of something. When you responded by discussing your own sexuality a bit, this scared me even more. I was getting very uncomfortable. I thought you noticed this, and I guess I felt that you should've said something about it.
>
> *Carol:* So, Steve, you really began to put on the brakes. You felt anxious about getting in over your head. How about you, Bill? You looked a bit agitated to me. I was wondering what was going on inside you.
>
> *Bill:* I had the feeling that you brought up the issue of sexuality because you thought it was an area *I* should look into. So I began to talk about it, and then you became just a listener. That made me a bit angry, because I thought I had been duped into doing something that I hadn't chosen to do. I stuck with it, though, because I wanted to show you that I can face up to issues when challenged to do so.

Note that each person discovered immediacy issues that they didn't face up to during the conversation itself. The purpose of this exercise is to sensitize the partners to such issues by having them "recall" them after the conversation is over, with a view to dealing with these issues *as they arise* rather than later in a special session. Steve and Bill's conversation would have been quite different had they had the skill of verbalizing their "immediacy awarenesses" to each other as they arose.

Exercise 49: Some Basic Immediacy Questions

From time to time you will ask yourself questions about the quality of your relationship with each of your fellow group members. The questions here need not all be answered each time you want to review your relationship with another person, but they do serve as a kind of checklist of some of the major factors in a relationship.

Directions

When you want to review a relationship with anyone in your group, read the following questions. Check those that have particular relevance for your relationship with that person. The other person should do the same. The list should provide concrete issues for discussion between you.

Immediacy Questions

1. Do I take the initiative to contact you with some frequency?
2. Does either of us take significantly more initiative?
3. If one of us takes significantly more initiative, how does each of us feel about this?
4. How readily do I disclose myself to you? In what ways do I entrust myself to you?
5. What do I tend to hold back from you?
6. Do I spend an adequate amount of time communicating understanding to you when you reveal yourself?
7. Are there any ways in which I feel phony when I'm relating to you?
8. Even though we are in a time-limited group situation, do I express genuine interest in you? How?
9. Do I challenge you? Am I interested in the demands you want to place on yourself? Am I willing to be a resource person for you in the interpersonal areas in which you want to change?
10. Do I challenge you through deeper levels of understanding (AE II)?
11. Do I give you straight feedback on your interpersonal style?
12. Do I confront you? What modes of confrontation (experiential, strength, weakness, didactic) do I use? In what areas am I a victim of the MUM effect with you?
13. Do I initiate immediacy issues with you? What issues between us do we tend to skirt or avoid?
14. Do I make unrealistic demands of you? If so, in what areas?
15. In the group, do we pair up together in order to protect each other or for any other purpose?
16. If we have a poor relationship, what does it lack? What negative or alienating elements are at work in it? Do we want to improve the relationship? Is it too much work? Too unrewarding? Are we too different?
17. If we have a bland or "plateau" relationship, is this what we want? Do we find each other boring? If so, what bores us? Is there any reason that we should raise our expectations of each other? What have we failed to do? Are we too much the same?
18. If we have a good relationship, what are we doing to maintain this high level? What kind of directionality do we want in our relationship?

Since each of you will be asking himself these questions, share your answers, focus on what you commonly agree are the most important issues, and determine how you want your behavior toward each other to change.

When you take up delicate areas, call on the resources of the group, asking other members for feedback. Ask them to help monitor new behavior you decide to use with each other.

Exercise 50: Summarizing as a Preparation for Immediacy Exchanges

In order to determine where you stand with another person right now (relationship immediacy), it is useful to review the highlights of the development of your relationship.

Directions

Picture each member of your group in your mind's eye and write out a short summary of the highlights in the development of your relationship. Write it as if you were talking directly to the other person.

Example

Norm, at first you were a bland person to me. You were rather quiet and, I thought, timid. I tended to ignore you, not paying too much attention to you even when you did speak. In my stupidity, I didn't think it important to relate well to you, because I didn't see you play an important enough role in the social structure of the group. That obviously says a lot about me. But, as I came to realize that we were a learning community and not just a microcosm of the world outside, I began hearing you as you spoke to others. You were "quiet" in some sense of the word, but hardly weak or timid. You listened to others and responded well and with feeling. I actually began comparing myself with you, seeing you as a better group member. I got down on myself. I still feel phony in comparison to you. There's something solid about you I find appealing. Recently I've wanted to talk to you more directly, but I'm a bit ashamed to do so because of the way I reacted to you in the beginning. You've made direct contact with me relatively little. I'm not sure why. One guess is that you didn't see me as very receptive. Obviously, you no longer appear bland to me; but, now that I can't categorize you or put you in the "proper" social place, I'm uneasy. I don't know how to relate to you.

Write out or speak out such a summary for each member of your group. It is most efficient to use the round robin format to exhange summaries. When the exchanges take place (they should be relatively short and concrete), you will find out whether each other sees the development of the relationship (or its nondevelopment) in a way similar to yours. In any case, you will establish a basis for discussing how each of you would like to change his behavior for the other.

CHAPTER 11: FURTHER READINGS

Carkhuff, R. R. *The art of helping: Trainer's guide.* Amherst, Mass.: Human Resource Development Press, 1975. See Chapter 5.

Gazda, G. M. Perceiving and responding with immediacy. Chapter 18 in *Human relations development.* Boston, Mass.: Allyn and Bacon, 1973. Pp. 148–154.

PART 5

Phase II: The Skills of Effective Group Participation

Chapter 12

Group-Specific Skills

INTRODUCTION

The three sets of skills I have discussed and illustrated so far are the building-blocks for high-level group participation. If you are adept in these skills, you have the capability of communicating effectively in a one-to-one situation. Having these skills, however, does not mean that you will automatically be an effective communicator in a group. To be an active, effective communicator in a group situation requires a specific set of skills. Group-specific skills are those mediating skills that enable you to translate basic interpersonal skills into effective group interaction.

My own experience with groups has made me take a couple of steps backward, as it were. Let me explain. My first experience as a participant was with unstructured groups—that is, human-relations-training groups that had no specific goals or contract. Some theorists (see Bradford, Gibb, & Benne, 1964) even suggest that such groups *should* be characterized by what they call "planned goallessness." In such groups I often saw the following process taking place. The members of the group would mill around quite a bit trying to decide what to do. The group leader generally provided no direction, but every once in a while he would comment on what we *were* doing ("It seems to me that there is a struggle for leadership here," or "The silences get longer after any particular member tries to initiate something or provide some structure but gets shot down"). That is, the leader assumed the role of *process observer.* Gradually we worked at establishing the ground rules for our interaction, finally elaborating some kind of contract or goal structure to give direction to the group. However, by the time we finished all that work, often little or no time remained to put into practice the contract we had so painfully elaborated. It seemed that we were always organizing ourselves. Although it is true that we probably learned much through this experiential process, many of us were dissatisfied that we never got the opportunity to move along further—to *do* what we had organized ourselves to do.

Therefore, I elaborated a structured approach to groups (Egan, 1970, 1973a) that translated a specific goal structure into a contract to which all members of the group would subscribe *before* entering the group experience. This goal structure added a great deal of directionality and cohesiveness to the groups. The participants didn't have to spend a great deal of energy trying to determine what to do. But something was still missing. It was evident that some of the members who subscribed, even eagerly, to the contract didn't fulfill it very effectively. Since I did not want to assume (although I was tempted to do so) that such failures were due to ill will, I looked further and discovered that, in many cases, the contract demanded more than the participant could deliver.

I had failed to appreciate the fact emphasized at the beginning of this book: that high-level communication in groups is complicated and demanding. Good basic interpersonal skills are called for. Some of those who failed to observe the contract did not have a solid foundation in these basic interpersonal skills. Thus I took my first step backward: I introduced a two-phase approach to human-relations training (Egan, 1973b), the first phase of which involved training in basic human-relations skills. Group participants were first trained in responding and challenging skills and then asked to involve themselves in a group experience governed by a contract to which all subscribed. This helped, but it was not the ultimate solution. Some people, even people who did well in the skills-training phase of the lab, failed to be good group members. Again, it was either a question of their ill will or my myopia. Fortunately, it proved to be my myopia again.

I found that group participants did not automatically transfer the skills learned in a one-to-one setting to a group setting. Another step backward (or was it a half-step forward?) was called for: training in group-specific skills, or the skills of using human-relations skills in a group setting. This chapter, then, is central: it describes what it means to be in a group precisely as a member of a group. One way of illustrating the difficulty of being a member is to point out the differences in the role behavior called for between practicing skills in a three- or four-member training subgroup and using these skills in an open-group experience. In the skills-training subgroups, you were asked to assume *one* of three roles at any given time: you were either speaker *or* respondent *or* observer. Your task was relatively easy because it was specified for you. However, in the group, it is now up to *you* to engage in all three sets of behaviors and to make the judgment when you should engage in each.

- *Speaker role:* you must initiate dialogue—for instance, by disclosing yourself or by taking up immediacy issues with your fellow group members.
- *Respondent role:* when others speak, you must take the initiative to respond to them—for instance, with primary-level accurate empathy.

- *Observer role:* you are to attend to all the interactions that take place in the group, give feedback, and "own into" the conversations taking place.

In the practice of basic interpersonal skills, as you move from one-to-one situations to the group you may experience what could be called "role overload" as the speaker-respondent-observer demands of group interaction hit you all at once. Perhaps *the* group-specific skill is putting all of these skills together into a free-flowing group conversation. This chapter will provide some practical ways to do this. Ultimately, however, communication is an art, not just a set of technical directives. The artistry of communication depends on the kind of person you are, and the kind of person you are depends on how you involve yourself in life.

USING SELF-DISCLOSURE SKILLS IN A GROUP SETTING

Most people find it more difficult to talk about themselves in a group setting than in a one-to-one situation. Indeed, when I suggest to some people that they may profit from involvement in a human-relations-training group or counseling group, they respond (at least nonverbally) as if I had asked them to take a trip to Antarctica. They don't feel prepared for a group experience because their lives have been devoid of similar experiences. Their educational experiences have centered around a teacher; perhaps they have been in an occasional "group discussion" of some type, but they have never been asked to be personal in a group or to share themselves with any degree of intimacy. What has your experience with groups been up to this moment? Where in your life has there been some form of group intimacy—in your family, or within your circle of friends?

"Let George Do It"

Paradoxically, people tend either to be more conservative in a group than they are in day-to-day living or to take greater risks than they usually do (see Shaw, 1971, pp. 73–79). You may be more conservative in a group than you ordinarily are if those who speak up most often in the group are also conservative. That is, you may be heavily influenced by the most vocal members of the group. Your degree of participation may also be a question of responsibility. When you're on your own, what happens stands or falls on what you and you alone do. However, in a group setting, since there are a number of members, it's too easy to fall into the habit of "letting George do it" or "letting the other person go first." Second, you may trust some members of the group more than others, and you may let the quality of your least trustworthy relationship set the level of your self-disclosure. The obvious solution here is the use of immediacy: you should work on relationships in which your trust is low. On the other hand, if you do have solid, trusting relationships with a number of the members of your group, you can learn to let these relationships determine your level of self-disclosure, for it is from these relationships that you receive encouragement, understanding, and support.

> *Group Member A:* John, as you know, you are relatively silent here. I don't want to belabor that point. It has been brought up a number of times already. I'd like to tell you, however, what it does to me. When you're silent, I don't know what you're thinking, and I begin to get fearful that you're judging me—that you see what I'm doing as not worthy of any kind of response. My tendency then is to become silent myself, or at least not talk about myself more deeply. I guess I see this as both your problem and mine. I want to count more on the encouragement and support I do get when I reveal myself. But I also want to know where I stand with you.

This kind of immediacy disclosure can pave the way for other forms of self-disclosure.

What Is Happening to Me Because of What Is Happening in the Group

Immediacy is an important form of group-specific self-disclosure. It differs from self-disclosure that is a direct revelation of some facet of your interpersonal style.

> *Group Member B:* I'm an impatient person. If I have something to say, I want to say it immediately even if it doesn't fit. Therefore, I sometimes interrupt conversations here to get on with my own agenda. I think all of this is related to my tendency to be pretty self-centered. When I want to take time for myself in this group, I just take it.

Here B is talking about his general interpersonal style and how that style affects his participation in the group. Notice how the following statement differs.

> *Group Member B:* I feel ashamed of myself right now. Karen, I know you're trying to work out the way you tend to distance yourself from George, but inside I'm impatient. I want the two of you to finish up your business quickly because, frankly, I want some "on" time. I want to talk about myself. I see now that I do this all the time. I've stopped thinking about your needs; I have to gratify my own.

Here B is disclosing what is going on inside himself in the here and now. It obviously relates to his interpersonal style, but it is more directly related to what is happening right now in the group. Too many group members think that what is happening "inside" is not important enough to bring up. Or they bring issues up only after they have become history (and therefore safe). Notice the difference between the following statements.

> *Group Member C:* Neil, last week I could tell that you were holding something back. I think you were really angry and were investing all your energy in holding your anger back. I sat there scared. I didn't know why you were so angry. I don't believe anyone confronted you irresponsibly, or anything like that.
>
> *Group Member D:* I'm very uncomfortable right now, Neil. You don't seem to be yourself. You're holding onto the side of the chair tightly. I think you're angry; but if you are, I don't know why. I'm afraid to ask. Something inside me is saying "Keep out of it." But there seems to be a kind of conspiracy of silence here—and it only makes me more uncomfortable.

Whereas C talks about his experience and feeling only from a safe historical perspective, D risks here-and-now disclosure of what is happening inside him. It happens with some frequency that group participants write very rich and engaging logs but fail to disclose what they write to the group.

Exercise 51: Encouraging "What-Is-Happening-to-Me-Right-Now" Self-Disclosure

This exercise can be used from time to time to tap some of the unused resources of the group. Its purpose is to get at the here-and-now experiencing of the participants. It may well be used when the group seems to be facing some kind of crisis or when "nothing is happening" in the group.

Directions

Stop the interaction (or interrupt the silence). Give the members of the group about two minutes or so of silence in order to reflect on their present internal experiencing. Each group member should ask himself "What have I been thinking or feeling or experiencing in the last half hour that I have *not* verbalized but that would contribute to the group?"

Some examples

> *Group Member A:* I've been bored. I didn't attend or listen. I actually shut myself off from the group and began to daydream about the weekend I'm going to have at the lake. I'm embarrassed to reveal this; I think I was really detracting from what was going on.
>
> *Group Member B:* John, I was getting angry at you because, although you were talking about yourself, you seemed to me to be engaging in a monologue.
>
> *John:* I was getting angry, too. I thought I was saying something significant about my style, but nobody was saying anything to me; nobody responded, and I felt out in left field. When there was no response I kept going, trying to explain more, and then began getting angry with myself for playing that game.

Note all the unused resources inherent in what these members failed to disclose. Failure to disclose leads to a boring, frustrating time for all.

Once the group participants have been given a chance to review their own experiencing, they can share it in a number of ways. For instance, members can take turns stating very briefly whatever disclosure they think may contribute something to the group interaction by stimulating further and more open interaction.

Getting help in self-disclosure. If you would like to disclose yourself in some way within the group but find it difficult to do so, use the resources of the group to help yourself. For instance, outside the group meeting, choose someone in the group whom you particularly trust. Tell him about your difficulty and ask him to help you disclose yourself within the group. Once you have told him, you will probably find it easier to tell others. You can also count on his encouragement and support as you try to do this in the full group. I am not implying that you should use this process in order to engage in "secret-dropping" within the group. I assume that what you are trying to reveal is appropriate and relevant to *this* group of people. You can check out the appropriateness of your disclosure with your resource person outside the group.

The use of there-and-then self-disclosure. One of the criteria for appropriate self-disclosure in a group experience is that the disclosure be relevant to *this* group of people in *this* situation. A human-relations-training group is not the same as a counseling group, where the focus may well be on problems outside the group. Its goals and contract define the human-relations-training group as very much a here-and-now process. Therefore, if you want to talk about your feelings, experiences, and behaviors that take place outside the group, it is essential that you relate these disclosures to these people in this situation.

One way of ensuring that your disclosures in the group are appropriate and that there-and-then disclosures are made relevant to the group is to *prepare* what you want to disclose about yourself. This means that between group sessions you give some thought to what you would like to disclose about yourself: "What could I disclose about myself that would help me move more deeply into community with my fellow group participants?" It goes without saying that if you then disclose yourself in a mechanistic way, it will be counterproductive. Preparing is not meant to cut down on the spontaneity of your self-disclosure in the give and take of the group experience itself. However, preparing some things you would like to say about yourself leads to a fuller group experience and paves the way for greater spontaneity of self-disclosure. The following exercise can help you in this preparation.

Exercise 52: Preparing Self-Disclosure

In the give and take of the group experience, sometimes it isn't easy to figure out what to say about yourself. This exercise gives you time to reflect on what self-revelations may help you accomplish your own goals while pursuing the goals of the group contract. Obviously, this exercise should serve as a stimulus for self-disclosure, not as a way of limiting yourself. It is also obvious that it is impossible to prepare for "what-is-happening-to-me-right-now-because-of-what-is-happening-in-the-group" self-disclosures.

Use the following form between sessions to help prepare or stimulate self-disclosure. An example of how this form may be used is included below.

Self-Disclosure Form for Members of a Human-Relations-Training Group

1. What would you like to disclose about yourself *at this time* in the development of the group that would enable you to pursue the goals of the group more effectively (such as dealing with your interpersonal style, establishing and developing relationships with the other participants)? Be as concrete as possible.

Examples

Group Member A: I want to mention, at least briefly, how poorly I relate to my father. My mother is dead, and my father and I are constantly hostile to each other. I can't please him at all. He disapproves of my major (history), saying it's useless, of my friends ("losers"), and of my part-time job (counseling in an agency that deals with mentally retarded children).

Group Member B: I feel very hypocritical about what I am doing. Within the group, my skills are excellent. I am caring, I attend, I show a great deal of respect, I challenge because I care. Outside the group, my behavior is very different. It's not that I do the opposite; it's that I don't take the initiative to put practically any of what I learn in the lab into practice. I know now that I have interpersonal skills, but I'm also aware that I can be lazy or scared or a confirmed loner outside the group. I guess none of these skills are values for me in everyday life.

2. Show how what you propose to disclose about yourself is relevant to *this* group at *this* time or to your relationship to any one member of the group. If your disclosure deals with there-and-then matters, legitimize it by relating it to the here and now of this group. Does your disclosure relate to the contractual goals and to your own particular goals within the group?

Group Member A: Peter [the trainer], James, and John are quite a bit older than I am. My interactions with them are sparse and overly reserved. I'm beginning to think that I see a bit of my father in *all* older males, especially if they have some position of authority (like Peter). I'm afraid of Peter, but I haven't told him so. I want to deal with my hesitancies with all of them.

Group Member B: I want the help of my fellow group members in order to find out whether I'm just playing a game or putting on a show for them. They don't see the day-to-day me. I want them to help me examine interpersonal relating as a value; that is, I would like to get others to share their values with me. I want to see how their everyday experience differs from the group experience. Are we all just "on" for the group and something else outside?

3. How do you want to reveal yourself? That is, do you want to address yourself to the whole group or to some particular group member or members? If the disclosure is difficult, do you want to "rehearse" it in some way outside the group with some group member you trust?

Group Member A: I'll talk to John first, because I have the best relationship with him among the three. I will explain my situation at home briefly and then tell him how I see it affecting him. Afterward I'll try to address James, and then Peter, about the same issue.

Group Member B: I'll make a statement to the whole group, because I think this is an issue that affects my relationship to everyone. I'd like my disclosure to be concrete enough and real enough to act as an indirect challenge for the others to share where they stand on this value issue.

Self-disclosure and intensity of living within the group. I have seen groups whose members had achieved a high degree of competence in the use of the three sets of basic interpersonal skills described in this book, but whose interaction—no matter how highly technically competent—could still be described only as bland. If you and your fellow group members are afraid to take reasonable risks in self-disclosure, then all the skills training you undertake in order to prepare yourself for the group experience is similar to setting a beautiful table with elegant linen and tableware and then serving hot dogs. As Shaw (1971, pp. 73–79) indicates, groups are capable of what is called the "risky shift"; that is, the members collectively will take risks that they wouldn't take as individuals. This willingness to take greater risks in a group setting is not completely understood, but evidence so far indicates that it can be attributed to a number of factors. Wallach and his associates (Wallach, Kogan, & Bem, 1964) suggest that risk-taking in groups is easier due to the diffusion of responsibility that takes place there; the responsibility for a risky decision can be spread out among a number of people. The individual group member doesn't feel the entire weight of responsibility. Second, risk-taking is a value associated with various roles in our society—for instance, the job of manager. The role of "high-level" group member in a human-relations-training group is also one of the roles in which risk-taking is approved of and rewarded. Third, risk-taking is generally a cultural value: "The general idea is that people in our society value risk, and in the group situation most individuals are willing to take risks in order to enhance their status in the group" (Shaw, 1971, p. 76). Finally, risk-taking individuals tend to be influential in groups. Or at least they serve as a kind of stimulus for others to take the risks they want to take anyway. Therefore, a member who discloses himself in a more than superficial way in the group (provided that his disclosure is reasonable and appropriate) is usually rewarded for "doing a good thing" and thus does tend to influence other participants to talk about themselves a bit more freely.

I am not suggesting that the "risky shift" phenomenon be used to manipulate others or to get participants to engage in secret-dropping and/or psychological nudity. But too often the self-disclosure in a group is too "thin"; it does not equal the quality of the disclosers' skills development. A certain level of self-disclosure is necessary if the group is to be an intensive experience. The group is more than a skills-building experience; it is also a relationship-building experience and an exploration of one's interpersonal style. Some relative depth of self-disclosure is part of the price of achieving these goals. According to Whitaker and Lieberman (1967), the most significant therapeutic experience is the discovery that feared consequences do not occur. Analogously, feared consequences in human-relations-training groups often do not occur either, and too many participants invest a great deal of energy in fear and defensiveness instead of in reasonable risk-taking and its rewards.

USING RESPONDING SKILLS IN A GROUP SETTING

As important as empathic understanding is in creating a climate of trust and support in the group, there is still a distinct tendency for group participants to fail to supply high levels of understanding within the group. As long as the structure calls for accurate empathy—precisely the case in the individual-skills-training phase of the laboratory—participants tend to learn the skill and do it well. However, once it is up to individual initiative to respond to other participants with accurate empathy, it becomes evident that supplying high levels of understanding is not a second-nature process with most people. *You will not tend to respond frequently with accurate empathic understanding within the group unless this understanding is part of your day-to-day interpersonal style.* Sometimes a group develops "specialists" in accurate empathy, and others then become lazy and leave this task to the specialists.

Therefore, you should both monitor your own behavior and ask for feedback from your fellow participants on the quantity and quality of your understanding. Your specializing in accurate empathy means that you are fearful of other kinds of interaction, such as self-disclosure, confrontation, and immediacy. If you are a specialist because you feel the need to devote your energy to raising the level of understanding within the group, you should bring this problem to the attention of the whole group and specifically to the attention of those who don't provide enough human nourishment through accurate empathy. Another reason why it is absolutely necessary to achieve relatively high levels of accurate empathy (coupled with genuineness and respect, expressed behaviorally) is that, through the communication of empathic understanding, you win the "credit" needed to intervene more strongly in the lives of your fellow group members—through confrontation and immediacy. Thus, if you don't achieve adequate levels of empathic understanding within the group, you cut down on your own freedom (you must restrict your interactions) and on your own ability to influence others (mutual social influence being a dimension of interdependence.)

Initiation and Accurate Empathy

If, as you watch yourself and get feedback from your fellow group members, you discover that you don't contribute sufficiently to the group in terms of providing accurate empathic understanding, how can you increase this necessary behavior?

Proceed mechanically. Make a contract with yourself to respond with accurate empathy a certain number of times during each group session. Certainly this is a mechanical way of proceeding, and not the ideal way, but it generally works; that is, it does help you increase the amount of human nourishment within the group. Just because your approach is mechanical doesn't mean that it need be phony or inept. The mechanical part is making sure that a sufficient *quantity* of understanding is present. As responding with empathic understanding becomes second nature to you, obviously you can drop the mechanics. The first steps in any learning process are often mechanical; this fact is part of being human. There are some people who don't want any part of the relating process ever to be mechanical. This attitude, I believe, is self-defeating when applied to learning and practicing human-relations skills.

Get a monitor. Tell someone outside the group that you are trying to increase this kind of behavior. Ask him to watch your performance closely, and get feedback from him after each session on both the quantity and the quality of your empathic understanding. Better yet, let the entire group know that you feel deficient in this area. Making your deficit public puts you under pressure to increase the desired behavior. If you are sincere in wanting help in monitoring your behavior, a number of your fellow members will surely give it to you.

Reward yourself. There may be other forms of behavior that you like to do in the group and do with some frequency—self-disclosure, challenge, immediacy. Don't allow yourself to engage in these behaviors until you have responded a certain number of times with accurate empathy. Then "reward" yourself for providing human nourishment by permitting yourself other kinds of behavior you like (provided that these behaviors are consonant with the overall contract).

Practice outside the group. Chapter 7, on responding skills, contained an exercise dealing with the use of accurate empathy in everyday life. If you want to increase this behavior within the group, increase it outside the group. Find out whether or not this behavior is part of your ordinary repertoire. If it is not, or if you seldom respond to others with accurate empathic understanding, find ways of increasing it. Make raising the level of empathic responding an interpersonal project.

Reexamine your skill level. One reason why you might fail to provide much empathic understanding is that you are relatively poor at this skill. If this is the case, pair up with someone else who is good at the skill, and learn it more thoroughly than you did in the practice sessions. Time invested here is time well spent, for you cannot move freely in the group without this skill. If you find that you are poor at the skill because your perception of what is happening in another's "world" is poor, you must sharpen your attending skills. If the deficit is in the area of the actual technology of the skill (the ability to *deliver* it by actually communicating understanding in a way the other can perceive it), you need more practice. Perhaps you could do written exercises again (or more carefully). Finally, if the deficit is in the assertiveness component of the skill (you are fearful or otherwise hesitant to use it), the suggestions above should help you help yourself increase the frequency of your use of the skill.

Lack of assertiveness may be related to your general lack of confidence or to low self-esteem. In these cases, there are two possibilities. First, use the group as a place where you can get some help with the problems of lack of self-confidence and low self-esteem. It is probable that other members of your group have similar problems, and therefore you won't become the group "helpee"; rather, a climate of mutual help will develop. Second, successful attempts at assertiveness—even relatively mild forms of assertiveness—do much to increase self-confidence and raise self-esteem. Building up self-confidence and self-esteem can be slow, painful processes; but being an initiator in the group, provided that there is an adequate level of trust and support in the group, can accelerate these processes.

Using Advanced Accurate Empathy and Confrontation in a Group Setting

These two skills are considered together because they are both "strong medicine" with the capability of inducing some kind of disorganization (preferably beneficial) in the person to whom they are communicated. The most common error in the use of these skills in a group setting is failure to use the resources of the group. For instance, A confronts B (we will assume that he does so responsibly), but the other members of the group are not then called upon, by either A or B, to offer their comments concerning the confrontation. In this instance, at least, A and B are trying to develop their relationship *in front of* the group (with the group as audience) instead of developing their relationship *in the context of* the group and through the resources of the group. For example, when A confronts B, he could end the confrontation something like this:

> *Group Member A:* At least this is the way I see your behavior. I'm not sure whether it's just me who sees you that way, or whether others do, too. I think it would be good for us to check this out with the group.

This statement can be worded in many different ways, but, whatever the wording, it constitutes a kind of group perception-checking. The purpose is not to gang up on the one being confronted but to make sure that he is not victimized by highly idiosyncratic perception. If A, for instance, finds that he is the only one who sees B in that way, the problem is clearly A's—or it has to do with the way A and B relate to each other. This leads to an immediacy question: how do A and B relate to each other? If A does not enlist the resources of the group to help with this question, B can do so.

> *Group Member B:* I'd appreciate finding out whether others in the group see my behavior in the same way. This is a touchy area for me, and I think I could use some help in exploring it.

The assumption here is that B first tries to understand what A is saying to him. Then he uses the resources of the group to clarify what A is saying and to explore it more fully. Obviously, B could use the remark above as a defense, but it could equally be what it is meant to be—a calling on the resources of the other group members.

I have already mentioned that self-confrontation is an ideal. Here I'd like to add that the person who confronts himself can call on the resources of the group directly to help him explore discrepancies, failure to use his strengths, and so forth.

> *Group Member C:* I think I have the skill necessary to initiate many more conversations than I do. Frankly, I think I'm lazy. I get away with it by giving all the signs that I am a kind of pleasant introvert. That way people don't expect me to take the first step. Does this ring a bell with any of you? I want to be more initiating here, and I'd like some help in placing that demand on myself.

One way to use group resources and to make confrontation more integral to the group as a learning community is to use any given confrontation as an opportunity for self-confrontation.

Group Member D (in reference to C's self-confrontation above): I'm not lazy, but I'm so fearful to initiate that the result is the same. I sit here safely waiting to be contacted by others. Even then I'm fearful that I might be contacted too strongly, or that the topic might be one that would cause me anxiety—like sexuality.

I don't mean to imply that D should respond to C by shifting the attention to himself. I am saying that both D and other members can use C's self-confrontation as an opportunity for examining themselves on the question of initiative in the group. This way, no particular member becomes a helpee while other members ring around him as helpers. In summary, *when any confrontation takes place, each member could well ask himself whether he should confront himself on that point.* Each confrontation, then, will be the presentation of a theme that all may consider.

In using both advanced accurate empathy and confrontation, remember that they should serve the purpose of relationship-building. Interpretations and confrontations that alienate tend to subvert the goals of the group. AE II and confrontation are, at their best, modes of involvement—ways of sharing life with another. AE II and confrontation are group-specific skills to the degree that they foster mutuality within the group.

Exercise 53: Confrontation Round Robin: Confronting and Responding to Confrontation

This exercise will help you "break the ice" with each of your fellow group members using confrontation and response to confrontation. It provides structure within which to take risks that you or others might otherwise avoid. It also provides structure to enable you to respond to confrontation nondefensively.

Directions

1. In your free time, write a few notes on a "strength confrontation" for each of your fellow group members. For each person, find a talent or resource that he or she is not using fully in the group and confront the person on that strength.
2. Let's call the partners A and B. Partner A in the round robin should (a) identify the strength, resource, or talent in question and (b) invite B to examine his use of that strength and how he might put that resource to better use within the group.
3. B should respond first using accurate empathy; that is, he should indicate to A that he understands what A is saying. Only then should B proceed to explore the areas suggested by the confronter.
4. Allow several minutes for a dialogue examination of the issue.
5. Stop and reverse roles. B is now the confronter, A the confrontee. Repeat the process.

Unless there is some reason for not doing so (such as limited time), every member of the training group should have a round with every other member. Thus the ice-breaking process will take place with each dyad.

Example

Partner A: In our group sessions, you take pains to see to it that there is a great deal of accurate empathic understanding going on. You yourself try to understand others, and you urge the other members of the group, principally by your example, to do the same. You're always genuine, and most of the time you're quite accurate. However, you tend to limit yourself to primary-level understanding. You're very slow to make demands on the members of the group, even though you're often in the best spot to do it—for example, by using advanced accurate empathy or confrontation. Your rapport is excellent, and I think you might be able to use it to help others make more demands on themselves.

Partner B: You see me as quite good at basic accurate empathy, but perhaps I'm becoming a kind of "specialist" at it to the exclusion of other, stronger kinds of group interaction. Since I do take such pains to understand, I gain the "credit" to intervene more strongly, but I don't take advantage of this resource. I should work on increasing my initiating skills.

Partner B then moves on to explore the content of his confrontation with A for a few minutes.

After both A and B have had a chance to confront and discuss, they should give each other feedback both on the quality of their confrontation and on the quality of their response to confrontation.

I do not suggest that each pair can completely work out the content of the confrontation in the time assigned. However, after the round robin is over, there are other possibilities. Each participant should review his "set" of confrontations and see whether there is any theme running through them—for example, lack of initiative. This theme can then be brought up before the entire group. Also, any member can review any of his confrontations in front of the entire group and use the resources of the group to explore it more fully.

Using Immediacy in a Group Setting

Immediacy is a cardinal group-specific skill; it is essential both to the process of examining your interpersonal style and to the process of establishing and developing relationships with the other members of the group. Immediacy—both relationship immediacy and here-and-now immediacy —offers rich opportunities for using the resources of the group. When A and B are discussing their overall relationship or trying to find out what is happening between them in a given here-and-now situation, they should call on the resources of the group for help in these processes, and other group members should spontaneously offer their views.

Sheila: Tony, outside the group we kid each other a great deal. Inside we are "polite" with each other; sometimes we're almost formal. I guess I see myself as being careful. I don't get serious with you, I don't make deeper contact with you. I avoid intimacy with you.

Tony: My experience is about the same. At one time I thought we just considered each other to be rather bland people—not worth investing in. Your "politeness" does make me feel bland. But this is a new wrinkle—avoiding intimacy. I need help on that one.

Marlene: My first impression was that you didn't care to relate to each other. But I see it differently now. Sheila, I see you as aggressive with the females in the group but much more cautious with the males. For instance, you've confronted me—and you did a good job of it—several times, but I don't think you've confronted George on anything. I guess I see your "blandness" with Tony as a kind of caution. Tony, as you know, you're relatively quiet here. When you do speak, often enough you do so with a smile and gentle humor. It's easy for me to like you, but I don't feel intimate with you. So maybe "avoiding intimacy" has something to it.

Here Marlene shares her impressions of both Sheila and Tony as these impressions relate to the immediacy issue facing Sheila and Tony. She becomes a resource for them. Any two members can deal as openly as possible with their relationship, but often it is very helpful to see themselves as others see them. Any given pair can be too close to their relationship to get a comprehensive view of it.

Group Member A: Yes, I see you relating to each other very closely. When you interact with each other, you do all the things the contract calls for—understanding, confronting, and the rest—and you do them very well. But, at least from my perspective, you seem to be in some sense "outside" the group when you relate to each other. I find myself at times feeling that I'm eavesdropping on a private conversation. Somehow you don't seem to bring your relationship and its richness into the group. Part of it is that neither of you seems to address others with the same enthusiasm. I think it's a question, not of making your own relationship anything less than it is, but of using the resources that you show in it in establishing other relationships. It may be that I just feel left out. I don't know how the others feel.

Here A tries to give these two other members his own perspective on their relationship. The method he uses is a strength confrontation.

Since immediacy usually implies some element of confrontation, care has to be taken that immediacy isn't used as an excuse for attacking another.

Group Member B: Frankly, Gary, I don't see our relationship going anywhere. You're too smug and self-content to get into much of a relationship with anyone. I think you're playing games here. Maybe you just want three hours' course credit, and that's it.

This has nothing to do with relationship-building. It is merely an attack on the part of an angry and frustrated group member. If either confrontation or immediacy turns into an attack, it is up to the whole group to handle it. The person attacked should be the first to respond, but he may need help.

> *Gary:* I think I *am* smug; but, frankly, I find it almost impossible to respond to what you just said. I get the feeling that perhaps your hostility toward me has been growing over the weeks and that now it's being dumped on me. Maybe it would be better to get the issue of my "smugness" into the whole group.

Gary doesn't dodge the issue, but he does call on the resources of the group to handle it. However, if Gary were to sit there in silence, not knowing what to say, it would be perfectly all right for another member to intervene—not to "save" Gary but to pursue the goals of the learning community.

> *Group Member C:* Gary, my perception is that you just got "blasted" and that you're not sure how to respond. I'm not sure whether you're angry or stunned. I think that the way you two relate to each other is an important issue in this group, but I feel a bit stunned myself right now.

Here, the resources of the group are brought to bear on a group issue.

Exercise 54: Group-Resource Questions on Immediacy

These are questions you might ask yourself as you watch your fellow group members deal with (or avoid) immediacy issues with one another.

The Questions and Some Sample Answers

1. How, concretely, do I see the two of you relating?

- You are direct and open with each other. Generally, if you have a sensitive issue to deal with in self-disclosure, you talk first to each other.
- You seem to trust your relationship, for you don't have to use a lot of qualifying phrases such as "it seems" and "maybe" when you talk to each other.
- Neither of you "specializes" in initiating with the other. You each seem free to contact the other as you wish—and this refers to all modalities of contact, including self-disclosure, confrontation, and immediacy.

2. Where do you two seem to be heading in your relationship?

- I see you growing in both feeling comfortable with each other and desiring to be with each other. My bet is that, if you had the time and opportunity, you would establish a friendship outside the group.
- I'm wondering whether you challenge each other enough. I guess I'm saying that I see you as very comfortable together and working well together. But I'm not sure whether you are ready to share life goals and to challenge each other on these.

3. How does your relationship affect me?

- I am stimulated by the initiative you take with each other. If I and others here were to take that much initiative, we'd all be fighting for "on" time in the group.
- I sometimes wish you could find ways of sharing what you have with each other more directly with me. I'm not sure how this could be done concretely.

4. What impact does your relationship have on the rest of the group?

- Your relationship enlivens the group. When you two talk to each other, I notice that almost everyone attends carefully. At those times, others seem to take more initiative also. Anyway, in some sense the group becomes more lively.
- If you talk too long to each other without anyone else's joining in, the conversation seems to get more and more intimate, and I feel like an eavesdropper. It would seem a good idea for the other group members to buy into your conversation and for you to widen the issues to include others.

LEADERSHIP: INITIATING SKILLS AS GROUP-SPECIFIC SKILLS

As you may have noticed, little or nothing has been said so far about leadership within the group. In a sense, however, this entire book is about *leadership functions*—the kinds of behaviors that make group experiences effective. The sooner these functions become diffused or distributed among *all* of the group participants, the better. Carkhuff (1974b), in elaborating the skills of program development, suggests a concrete and useful leadership schema that can be applied to degrees of initiating (or not initiating) in human-relations-training groups. Initiators and noninitiators can be placed on a continuum of types: the Detractor, the Observer, the Participant, the Contributor, and the Leader.

The Detractor

The detractor not only does not contribute anything to the overall goals of the group but is actually an obstacle. To paraphrase a statement from the New Testament, not only does he not enter, but he prevents others from entering. The Detractor may participate in the group, but only destructively. What are some typical Detractor behaviors?

The treatment end of the continuum. Participants in human-relations-training groups have problems, as do all "normal" people. However, these problems should not be "front and center" in the training group. Suppose people can be placed along a continuum such as the following.

In need of treatment ———————————————— Ready for training

Mode of participation

Potential participants who are too far down toward the "in need of treatment" end of the continuum often become Detractors, even in spite of themselves, in human-relations-training groups. These groups assume, as we have noted, that the participants are not overly preoccupied with deficiency needs. It is undeniable that the human-relations-training model presented here is therapeutic or quasi-therapeutic; but if emphasis is placed on therapy the group becomes too heterogeneous, and the whole group moves forward at the pace of its slowest member. If this model is to be used as a form of treatment (see Egan, 1975b), it would be better to use it with a treatment population—a relatively homogeneous group. Sometimes a well-meaning person in need of treatment ends up in a group of people at the "ready for training" end of the continuum and decompensates even further, because he comes to feel that he is a misfit, that others are learning more quickly than he, and that he is an obstacle to others' growth. In such a case, it is not that the human-relations-training experience is destructive in itself but that the "fit" among trainee, fellow trainees, and methodology is poor.

Cynicism. This form of negativism is extremely destructive in groups. Perhaps the most damaging failure of the cynic is his refusal to initiate anything. To initiate is to show interest, and he can't be caught showing interest in an enterprise that is in some way beneath him. He waits until someone else initiates something and then sits in judgment on what is happening.

> *Group Member A:* Marge has had a pretty hard time of it this evening, Larry. You seem to be unaffected, almost distant.
>
> *Larry:* I can't get excited every time a female cries. It's okay if that's what she wants to do. It's not as if this group were real life.

The cynic usually hangs like an albatross around the neck of the group.

No need for skills training. The person who wants a group experience but doesn't want the individual-skills training as a preparatory stage usually suggests that he doesn't need skills training. In my own experience I have found that, on closer inspection, most who thought themselves already skilled had a misconception of their skills level. If such a person does go through skills training because he "has to," he too can hang heavy around the neck of the group, because he is inept in his transactions. He is a Detractor in a twofold sense. First, he interferes with the skills-training process for the rest of the group members, because his heart is not in it. Second, he arrives at the open-group experience devoid of essential skills. I have suggested in the first chapter a certain hierarchy of attitudes toward the skills-training phase of the lab. All can use Phase I as a skills check-up. If you find that you don't possess some of the skills, use the time to acquire them. If you see that you have some skills but also certain deficits, use the time to acquire those skills you need and to perfect those you have. If you feel that you already have the skills, use the time to acquire a skills-training methodology and become a resource person in the training of others. These are creative ways of approaching Phase I. The Detractor is not a creative person. He usually wants things done for him, and then he complains because they aren't done right.

This is hardly an exhaustive list of Detractor behaviors. Can you think of behaviors that can turn the training experience into a negative one—behaviors that prevent learning and the formation of a learning community? It is hoped that no "pure form" Detractor will be found in your group. Detractors are best eliminated before the lab begins or before Phase I is over. However, some of us engage in Detractor-like behavior from time to time. Obviously, such behavior is worth confronting whenever it appears.

The Observer

The person who does not reach out actively to make contact with other group members and who, when contacted, responds inadequately is an Observer—a noninitiator. Like the Detractor, such a person is a liability to the group. Ideally, potential Observers should be screened out before the group begins, but that is not always the case. Experience shows that Observers can make their way through the individual-skills training of Phase I but then become inert within the group itself. The structure of Phase I apparently carries them along, but when they must choose to put what they have learned into practice they fail. Unfortunately, too many Observers seem to make it through the selection process. Again, there are not too many "pure form" Observers in groups, but often too many group members too often subside passively into the Observer role. I tend to keep groups small—six members plus trainer and co-trainer. If groups are too large, the opportunity to sit back and become an Observer is too tempting for some. But the idea of the silent group member who is "learning a lot by just observing" is worse than nonsense. Observers may not be Detractors, but they still hang like millstones around the neck of the group.

Anyone who shows signs of becoming an Observer in the group should be helped to examine his behavior very early in the life of the laboratory. If the other group members don't make legitimate demands on him for participation right from the beginning, they are doing both him and themselves a disservice. It goes without saying that the fears and hesitancies of the potential Observer should be understood, but it also goes without saying that he should be helped to face them. One way of doing this has already been suggested: all the members of the group can talk about hesitancies and fears, each about his own. Thus the potential Observer will not simply become the group helpee but will see his own hesitancies in the wider context of the group. The potential Observer may still need more support and encouragement than the next member, but he will not sink into the mire of the Observer role. Once a group member becomes identified as an Observer, it seems almost impossible for him to free himself from the role.

The Participant

Whereas Detractors and Observers subtract from the group experience in some significant way, Participants do not. The Participant responds in some positive ways to the structure provided. For instance, he does not disparage, ignore, or merely submit to the individual-skills training of Phase I. Rather, he cooperates with the training program offered and tries to increase his repertoire and raise the level of his skills. Within the group the Participant is one who cooperates and responds. For instance, when invited to contribute through self-disclosure, he does so. When confronted, he responds to the invitation by exploring his behavior within the group. The Participant responds not reluctantly but willingly, and he displays a certain degree of initiative in his response, but he still cannot be called an initiator. At worst, this kind of responder is overly dependent on the initiative of other group members and runs the danger of living a kind of parasitic existence within the group. If the group has too many mere Participants, it will probably be a relatively dull group. As with all types, the "pure form" Participant is not found with great frequency in laboratory groups. However, there is again a real danger that too many members will allow themselves to be mere Participants too much of the time. Initiation, after all, demands not only a great deal of work but also a fair amount of risk. It is easy for a number of participants in a training group to slip for relatively long periods into the comfort of the passive-participation role. The passive Participant seems to say, at least nonverbally, "I'm here, all right; I'm even attending; if contacted, I will exert some effort in responding, I'll cooperate at least in some minimal way; but don't expect much from me in the way of initiative."

The Contributor

The hallmark of the Contributor in groups is initiative. The Contributor doesn't wait to be contacted by others but actively initiates interactions with his fellow group members. If he is in a six-person group, he realizes that 30 different relationships must be considered in the process of relationship-building (or *301* potential relationships if he includes all possible coalition formations!), so he gets to work on his own relationship-building process. He wants to make significant contact with each of the other five members, and not just wait for significant contact from them. If he is to be interdependent with Member A, he must both take the initative in contacting A and expect A to do the same. *The Contributors make groups go.* The Contributor possesses at least minimal amounts of the three sets of basic interpersonal skills and also the assertiveness and courage needed to put these skills to good use. In a group in which all of the members are Contributors, they have to fight for "on" time, and such a group is an exciting place to be. If your group has frequent "low" periods, there are probably too few Contributors in it or too many participants who too easily slip from the Contributor to the Participant or even the Observer role.

The Contributor is active even outside the group. He keeps a log, not just as a repository of his musings and reflections but as a tool for determining his agenda for the next group meeting. He comes to the group not to see what will happen but to make things happen; he comes with business to accomplish.

> *Group Member A:* Jan, I've thought about our relationship within the group this week, and, frankly, I come with some confusion. I'm physically attracted to you, and I believe that this attraction makes me cautious. I've begun to see that I don't know what to do with physical attraction. I let it get in the way. I suspect my motives in making contact with you. As you know, I don't initiate much with you at all.

Member A comes to the group with an agenda. His self-disclosure deals with a facet of his interpersonal style and its impact on relationship-building with another group member.

What kinds of initiation does the Contributor attempt within the group?

Human nourishment. The Contributor provides high levels of human nourishment in the group, especially through communicating primary-level accurate empathy. Since the communication of empathy is second nature to him, he initiates it consistently and genuinely. Providing understanding is always a part of his agenda.

Self-disclosure. The Contributor sees reasonable self-disclosure as a way of establishing trust within the group. Therefore, he doesn't wait until trust conditions in the group are optimal before revealing himself. If understanding is not forthcoming from his fellow group members, he faces that isssue directly with them.

Challenge. Since the Contributor both reveals himself and provides high levels of human nourishment, he is in the best position to challenge others. He is not afraid to do so. I have frequently asked group participants to reveal in what ways they were disappointed in their group experience. One of the most frequent replies is "I wasn't confronted or challenged enough." Most would also have to say "I didn't confront or challenge enough." Perhaps some participants do not challenge their fellow group participants because they realize instinctively that they have not acquired enough "credit" to do so. The Contributor has both the credit and the courage to challenge, and he does so caringly and responsibly.

Calling for specific feedback. Since the Contributor wants to grow, he doesn't merely sit back and hope to receive feedback on his interpersonal style. If it isn't forthcoming within a reasonable time, he asks for it.

> *Group Member A:* I've noticed recently that, at least relatively, I do a great deal of challenging around here. I don't want to become the group "specialist" in challenging. But no one says much about that part of my behavior. I'm not sure whether or not I'm upsetting people. I'm not always sure that I have the "credit" with each person I challenge in some way. Paul, I was challenging you earlier for being cynical about yourself and your potential in relating to others, and I'm uneasy about how that sits with you.

Without being a prima donna, he asks for the feedback that perhaps should be given more freely. If a group had only one Contributor, he or she would certainly seem to be a prima donna to the Participants, Observers, and Detractors. A Contributor among Contributors, however, is sensitive to time limitations and the needs of others. He will work to get his share of "on" time in the group and feedback, and do so without apology, but he won't place himself in center stage all the time.

"Owning" or "buying into" the interaction of others. In a group setting there are no private conversations, and the Contributor is well aware of this fact. Observers and even Participants often refuse to join ongoing conversations because they don't want to "interrupt" a dialogue. The Contributor, however, attends carefully to all the interactions of the group and asks himself constantly whether he has something to contribute to an ongoing conversation, or whether it relates to him in some way. The Contributor can put the following distinctions into practice.

- *Disruption.* I can "barge into" a conversation without helping in any way. I merely draw attention to myself or otherwise sidetrack the issues being discussed by A and B. In my own experience with groups, disruptions take place infrequently, because most participants are so fearful of interrupting that they don't even participate when participation is called for.
- *Interjection.* Interjection is a way of letting A and B know that I am attending; it can also be a way of letting them know, briefly, where I stand on a certain issue.

> *Group Member A:* Tom, I'm reluctant to talk to you right now. The last time we talked, you shifted the focus from your fear of strong women to my need for control. After the group session was over, I got really angry at you. Now that we're talking about the topics we're afraid to talk about here, I'm afraid you may shift the focus on me again.
>
> *You:* Yeah, I'm afraid even to tell you what I'm afraid to discuss!

Your statement here is an interjection. It states briefly where you stand in relation to the issue being discussed, and it indicates your attention and interest. An interjection can be a sign that you want to contribute and not just observe.

- *Involvement.* Involvement means that I feel free to join any ongoing conversation if I feel that I have something to contribute and if I determine that what I'm going to say is not a disruption. I take seriously the ground rule that there are no private conversations in the group. I can contribute to an ongoing conversation by

- communicating accurate empathy to either or both participants,
- sharing myself if it relates to what A and B are talking about and helps move their conversation toward their goal,
- confronting, if appropriate, but with care and involvement: "Tom, your slouching position makes it hard for me to believe that you're interested in what Bill is saying. And, Bill, you're not even looking at Tom. It sounds as if you're reciting a prepared list of grievances, and Tom's nonverbal message to you is 'Get lost.' "
- using immediacy: "Trudy, in no way do I want to gang up on you now, but I'd like to confirm what Jenny just said to you. When you become passive—go into what you call your 'little girl' behavior—I don't want to cuddle you, either. I want to shake you."

Since the cultural norm seems to be "Don't interrupt an ongoing conversation" (a convenient norm for the Observer), entering an ongoing conversation demands a great deal of assertiveness on your part. You may well need practice.

Exercise 55: Practicing "Owning" or "Buying into" Conversations

This exercise is designed to help free you from the tyranny of the cultural norm forbidding entering ongoing conversations. It provides enough structure and "permission" to enable you to practice.

Directions

The group should be subdivided into groups of three, Members A, B, and C.

A and B should begin a conversation (on some topic related to the group's goals—such as the quality of their relationship). They should talk for about five minutes.

During this time C should attend to what they are saying very carefully but should not enter into the conversation. During this time C should ask himself how he would like to intervene, what he has to offer that would help A and B get where they are going, what C himself could get from the conversation by entering it, and what resources he can bring to it in terms of understanding, challenging, and so forth.

After five minutes, C should enter the conversation in some way and "buy into" it for the next five minutes.

After ten minutes, the process can be repeated, with A and B also taking turns "owning into" the conversation.

This exercise can be used for those who have difficulty entering ongoing conversations. The group can even have a regular group meeting while the one or two people who have particular difficulty buying into conversations are assigned the role of "interrupters."

The "interrupters" should be given feedback on how they integrate themselves into the conversation in terms of disruption, interjection, or involvement.

Mutuality

The Contributor tries to foster give and take within the group. Since the Contributor engages in many kinds of initiating, he may give the impression of a virtuoso performer. If there is only one Contributor in the group or if the group trainer is identified as "the" Contributor, this virtuosity will be a clear danger. The Contributor should not contribute in ways that rob others of initiative. Immediacy—both relationship and here-and-now immediacy—is a primary concern for him; but all relationships, not just his, are important.

Being a Contributor is a goal for everyone in the group. Ordinarily, Contributors are made, not born; one becomes a Contributor by working hard. If you are presently a Participant, your goal should be to increase the quantity and the quality of your Contributor behaviors—gradually perhaps, but steadily. As Bion (1961) notes, it is impossible to do nothing in a group, even by doing nothing. A refusal to accept the challenge of agency in a group is a refusal to face one's potential. Erikson (1964) says of the agency/passivity dimension of psychic life:

> *Patiens,* then, would denote a state of being exposed from within or from without to superior forces which cannot be overcome without prolonged patience or energetic redeeming help; while *agens* connotes an inner state of being unbroken in initiative and in acting in the service of a cause which sanctions this initiative. You will see immediately that the state of *agens* is what all clients, or patients, in groups or alone, are groping for [Erikson, 1964, p. 87].

Agency or initiative, like independence, cannot be conferred; it must be seized.

> We can only say that [independence] has become manifest in an individual or group when it no longer occurs to that group or individual to seek the solution of its problems by an agent outside itself. To "demand one's independence" . . . is of course a contradiction in terms [Slater, 1966, p. 150].

The Contributor seizes initiative in the group, for he knows that no one, not even the trainer, is going to confer it on him.

The Leader

Carkhuff's (1974b) fifth mode of investment in an enterprise is called leadership. Perhaps here is the best place to discuss the function of the trainer in the human-relations-training group, for the trainer is expected to be a

Leader. In the phases that stress training in basic interpersonal skills, the trainer exercises leadership in a variety of ways. He explains the theory underlying the skills and demonstrates them, provides the training methodology, evaluates the performance of the participants, and teaches the participants how to evaluate one another. He both models the skills and provides directionality in the members' progress in acquiring the skills. He both supports and challenges the participants in their efforts to acquire these skills.

In the group experience itself, the Leader or trainer has several functions. First, he is a Contributor with all that that role implies. Perhaps the word "modeling" is not the best word to use to describe the trainer's behavior, for that word can have a patronizing tone. The trainer hardly manifests a "look-at-me" approach to the use of skills. But he does use his skills and use them effectively. In my opinion, unless the trainer is a Contributor he cannot provide other leadership functions.

Second, the trainer encourages a diffusion of leadership in the group; that is, he challenges and encourages others to become Contributors. In a way, he works himself out of a job; for, as more and more members become Contributors, less and less structure is needed in the group. If a member has hesitations about becoming a full Contributor in the group, one way he can avoid it is to idealize the trainer as *the* Contributor. This is an immediacy issue that should be faced early in the group by anyone who suspects that he is idealizing the trainer. If you feel yourself reluctant to interact with the trainer, or if you see yourself waiting for the trainer to initiate in the many little crisis points in the group, these may be signs that your expectations of the trainer are not realistic.

Third, the trainer is largely responsible for the directionality of the group, especially in its early phases. For instance, he implements the methodology for training in basic skills, assigns exercises, and establishes timetables in keeping with the needs and resources of the group members. He tries to keep the group moving at a pace (for example, in acquiring basic skills) that makes individual members "stretch" a bit. Sometimes an individual member will fall behind the rest (for example, he may be still struggling with AE I as the others are ready to move on to AE II). In this case the trainer assigns further exercises and calls on the resources of the group to help the slower member (for example, asks a member who is good at AE I to give some time outside the group to the member who needs some help).

The trainer is a Leader in that he is, fourth, constantly rethinking the training methodology, staying in touch with what the best in research has to offer, doing research himself, and using all of these resources to establish more effective training programs. A Leader is a high-level Contributor; that is, he contributes not only to the effective running of any individual group but also to the art and science of human-relations training. He reads the literature in the human-relations-training field hungrily, looking for ways to improve the training methodology and ways of expanding the use of human-relations training. For instance, some trainers who have worked with me have gone on to adapt the methodology outlined in this book to grammar-school and high-school populations. Indeed, most of the skills outlined here *should* be taught, at least in an adapted form, to children in the school systems (if not at home). There is something almost remedial about these skills administered at the college and graduate-school level. However, grammar- and high-school programs are just being started, and not on a very wide scale. The Leader is a person who sees the necessity for systematic training in a wide range of life skills and is capable of elaborating methodologies for training in these skills. Carkhuff, his associates, and those who have followed his lead have elaborated a wide variety of training materials in helping and effective-living skills (for instance, Aspy, 1972; Berenson & Mitchell, 1974; Carkhuff, 1973a, 1973b, 1974a, 1974b, 1975; Carkhuff & Pierce, 1975; Carkhuff, Pierce, Friel, & Willis, 1975; Egan, 1975a, 1975b; Friel & Carkhuff, 1974a, 1974b; Gazda, 1973).

In systematic approaches to human-relations training, the trainer does not apologize for structure or directionality. He proposes a methodology and expects the cooperation of the participants especially in learning the basic human-relations skills. Vince Lombardi, the late coach of the Green Bay Packers football team, was asked the secret of his team's success. He replied that his team was excellent in basic skills. For instance, they practiced long and hard until they could block and tackle flawlessly. I imagine that he and his assistant coaches didn't have to ask for volunteers to tackle the dummy. The same holds true for the trainer in human-relations-training groups. Although he or she is understanding, caring, and supportive, he or she is also challenging and demanding. Only trainers who are themselves Contributors can make such demands of others.

SOCIAL INFLUENCE

Most of us like to see ourselves as relatively free and independent people. However, one goal of human-relations training is to come to a deeper appreciation of *interdependence.* Interdependence does not involve a wholesale surrender of one's autonomy (dependence) or a continual fight to salvage one's independence (counterdependence). Interdependence means that people who are in possession of themselves and have a sense of identity feel free to surrender some of their autonomy in the give and take of their relationships.

Interdependent people admit that they influence one another. As soon as you involve yourself with your fellow human beings, you become both one who influences and one who is influenced. This give and take is called social influence (see Berscheid & Walster, 1969; Gergen, 1969; Kelman, 1967; Zimbardo & Ebbesen, 1970). I influence others both by acting and by not acting. For example, when I show care for others, they are influenced to like me, respect me, cooperate with me; when I am cynical, others are influenced to avoid and fear me. My silence at a meeting influences other members to think of me as impotent or unconcerned or forces them to deal with my silence; my failure to communicate understanding of others from their frame of reference in a group influences others to wonder whether I am "for" them and whether they can trust me.

From one point of view, the group experience is a social-influence process. Once we admit the fact that we do influence others and are influenced by them, we can make social influence an overt and constructive process rather than an unknown and potentially destructive and manipulative force in the life of the group and in our lives outside the group. In a sense, then, this human-relations experience is a kind of laboratory where we can study social-influence processes in the most pragmatic way—at first hand.

For a number of reasons, increasing your interpersonal skills can make you less vulnerable to random social influence. Skills training gives you a greater sense of competence and increases your self-esteem. You become less dependent and you are freed, at least to a degree, from the need for social approval. You also acquire the ability to challenge untoward attempts to influence you. At the same time, learning communication skills can open you up to more reasonable kinds of social influence. You can listen more carefully and with greater understanding to what others have to say. You are less defensive and therefore more willing to listen. All of this opens you to influence that fits in with your own value system.

A LEARNING COMMUNITY AND A MODELING NETWORK

Many current educational practices lead to competition rather than cooperation in learning. For instance, since medical schools accept only a fraction of students who enter premedical undergraduate programs across the country, the competition in such programs is often fierce. In this laboratory experience, however, you are urged to adopt a cooperative rather than a competitive stance. One way of looking at cooperation is to see yourself as one of a group of people who are trying to forge themselves into a learning community. In such a community, each member tries to contribute to building an atmosphere in which learning can take place most readily. Given the nature of the human-relations-training group, this means an atmosphere of

- caring, respect, mutuality, encouragement, positive reinforcement, support;
- seriousness, dedication, investment, availability, work, commitment; and
- reasonable challenge, directionality, and self-instituted and self-directed change.

In such a group laziness, sloth, cynicism, defensiveness, competition, indifference, and detachment are the enemies.

One of the assumptions of a learning community is that the members of that community care for one another. Caring includes the ability to get out of myself, to make my own concerns secondary for the time being, and to be with another in his world. Self-discipline and caring have provided essential bases for successful communities throughout history (see Kanter, 1972). Likewise, self-centeredness and excessive individualism have always spelled the end of community (Veysey, 1974). These rules apply even to such *ad hoc* communities as laboratory groups.

Ideally, your group will become a modeling network—one way in which social-influence processes can be used quite constructively. Practically, this means that you will be more likely to notice people who do certain things well—who have interpersonal talents, skills, and qualities that you would like to develop in yourself. People in the group who possess these qualities that you lack (for example, a kind of gentle assertiveness in interpersonal situations) can become "models" for you. Without losing your own identity, becoming dependent, or giving in to slavish imitation, you can begin to incorporate these skills and qualities into your interpersonal style in your own way. And the models in your group, once they become aware of what you are looking for, can become resource people for you. Obviously, you too can become a model of certain qualities for some of the other members of the group. The group as a modeling network is an example of a social-influence process and constitutes one dimension of the group as a learning community.

THE FISHBOWL OR THE MODIFIED FISHBOWL AS A PHASE-II TRAINING TECHNIQUE

How can you use the resources of the group to put into practice the group-specific skills outlined in this chapter? The most useful training technique I have discovered so far is the "fishbowl" or the "modified fishbowl." In Phase I you learned and practiced basic human-relations or interpersonal skills. In Phase II your task is to put these basic building blocks together in effective group interaction. You no longer have the relative safety of the dyad, the one-to-one communication situation. As I've mentioned before, group communication is much more complicated and much more demanding.

> As the number of people increases arithmetically beyond two people, the number of possible relationships increases geometrically . . . .
>
> It is no wonder, then, that as the number of people increase, the situation becomes considerably more complicated than any relationship between two individuals. Each individual has more relationships to maintain in a given amount of time. And it is not surprising . . . that many neurotics and psychotics are able to maintain one-to-one situations but "are unable to consider multiple relationships simultaneously." Such an array of possibilities is difficult enough for a "well-adjusted" person to comprehend. . . .

> Leadership emerges because it can reduce psychologically complicated relationships [Wilmot, 1975, pp. 18–19. References omitted].

Wilmot gives some indication why people at the "treatment" rather than at the "training" end of the continuum have special difficulty in Phase II: having to cope with multiple relationships simultaneously adds further disorganization to already disorganized lives. Second, Wilmot suggests that, in group situations, leadership (structure, directionality) is needed to cope effectively with "psychologically complicated relationships." The contract provides a great deal of directionality through its goal structure, and, indeed, one of the functions of the trainer is to be a kind of guardian or promoter of the contract—not in order to exercise control over the lives of the participants but to make sure that there is sufficient structure and directionality to make a relatively complicated communication experience viable and growthful.

The fishbowl (or modified fishbowl) is a structure that marshals the resources of participants individually and collectively and directs them toward the fulfillment of the contract.

Exercise 56: The Fishbowl

Format

The "fishbowl" always consists of two groups, but these groups may be constituted in different ways.

1. *The two-group fishbowl*

   a. *Two separate training groups.* One way is to set up two separate training groups, each with its own trainer (or co-trainers). Let's call the groups A and B. Let's also suppose that each of these training groups has six members, plus trainer or co-trainers. During Phase I, these two groups did not interact.

   b. *One group divided into two groups.* Or we can divide one larger group into two smaller groups. Let's say that we have one group of ten participants, plus co-trainers. In the fishbowl there will be two groups, A and B—each with five members plus one of the co-trainers. During Phase I this group divided into various subgroups in order to learn and practice the basic interpersonal skills.

In either case, each member of Group A is assigned a partner from Group B. The As form a group and interact together, and the Bs form a group and interact together. When the As are interacting, they form a group on the inside (the inside circle), while the Bs form a circle around them (the outside circle). While the members of the inner circle are interacting, their partners in the outer circle are silent, observing how their partners are operating within their group.

2. *The modified fishbowl*

Let's say that we have one training group of six members, plus trainer or co-trainers. If this group were divided into two sections, the separate groups would be quite small—perhaps too small for good group interaction. Therefore, partners are assigned, but the group never subdivides into two separate groups. In such a group you have two functions—first, to interact as you would normally do with your fellow participants and, second, to observe the behavior of your partner in a special way (as he will be observing yours). There are no "inside" and "outside" groups, but the partners will help each other in ways specified below.

Planning

Whatever the format used, partners meet for five or six minutes *before* the group meeting in order to review what they want to accomplish during the group session. Ideally, all the members are keeping logs, and between sessions they use these logs to develop a flexible agenda for the next meeting. Let's personalize this. Before you start your group meeting, meet with your partner. Since you are keeping a log and using it to formulate an agenda for yourself from meeting to meeting, you already have some idea of what you want to accomplish in the upcoming meeting. But first meet with your partner and use him as a resource to help you review what you're planning to do. *Partners act as resource people for each other; they help each other review agendas.* In the modified fishbowl a few more minutes must be allowed for the pregroup meeting, since both partners have to review agendas. In the two-group fishbowl, only the As plan their agenda, for they will be the first to constitute the inside (interacting) group. The Bs will get help from their partners later, just before they move in to constitute the inside (interacting) group. Below are some questions that you can consider with the help of your partner in determining and reviewing your agenda. Obviously, all of the following questions need not be dealt with in every reviewing session. Your agenda should be *concrete* and *manageable.* If your agenda is too long or too vague, you will always leave the group session feeling defeated. Concrete agendas are the fruit of concrete logs. Perhaps your agenda should contain one or two more items than you can manage to fit in during the group meeting, for this will serve as a kind of gentle pressure for you to get your work done; but the agenda should always be realistic. Some questions:

Self-disclosure should always be dealt with in the reviewing session, for it is one of the principal ways of "moving into the group."

- What is appropriate to disclose about myself? What disclosures will help me explore my interpersonal style and establish and develop relationships?
- To whom especially should I disclose something—for instance, ways I feel about that person?

- How good am I at disclosing what is happening to me in the here and now of the group?
- How can I make disclosures about there-and-then events and experiences and behaviors relevant to the here and now of this group?
- What have I written recently in my log that I haven't brought up in the group?
- What topics am I afraid to talk about in the group?

Responding is another principal way to "move into the group" and to gain trust and credibility with my fellow participants.

- How satisfied am I with my present level of responding to others, especially with responding with empathic understanding?
- Whom do I fail to understand in the group?
- To whom am I afraid to respond in the group?
- Is there enough human nourishment in the group in terms of empathic understanding?

Alternative frames of reference (AE II) refers to helping another see himself or others or the world from a different (more constructive) point of view.

- Whom do I understand well enough to present an alternative point of view to?
- Who could benefit from an alternative point of view? Am I the one to provide it?
- How could I concretely present such an alternative framework (for example, by summarizing)?

Confrontation is a form of risk-taking that pays off only if it is prepared for. It should also be invitational.

- Whom would I like to challenge?
- Can the challenge be stated in the form of a strength confrontation?
- Am I confronting behavior rather than motivation or internal psychodynamics?
- With respect to the individual I am about to confront, do I merit the privilege—have I developed a sufficient relationship with that person?
- What are my motives for confronting?
- How can I confront this person concretely? How can I describe the behavior (or lack of behavior) in question?
- How does this person's behavior affect me? The other members of the group?
- Have I legitimized myself enough through self-disclosure and understanding to be a confronter in this group?

Immediacy: Obviously I cannot prepare for here-and-now immediacy, but I can prepare for relationship immediacy—perhaps the best preparation for the more spontaneous here-and-now immediacy.

- What immediacy issues do I have with whom?
- Which of my relationships here are drifting? How can I contribute to greater directionality in these relationships?
- With which members do I cooperate? With which do I fail to cooperate?
- Are there any members I haven't taken time to understand?
- What immediacy issues have I been avoiding? What concrete experiences, feelings, and behaviors are involved?

Observation

In the two-group fishbowl, the As move into the center and interact for a specified amount of time to be determined by the trainer (sometimes shorter periods of 20 minutes or so put the members under a certain amount of pressure to accomplish something). Each member tries to accomplish his agenda or part of his agenda, but without trying to compete with his fellow participants. The Bs are in the outer circle observing the interaction. Each B observes the total group interaction, but especially the transactions of his or her partner. Use paper and pencil to jot down what your partner does and how well he does it. Observation is an important skill in itself—a discrimination skill. It is complicated by the fact that you are watching both your partner and the total interaction of the group. Use some shorthand to indicate what kind of skill your partner uses each time he or she interacts (for example, AE I, S-D for self-disclosure, a ? if he asks a question). Make some notation *each time* your partner interacts, so that by the end of the session you will have enough data to get an overview of his or her interpersonal style (the kinds of skills used and the frequency of each) and the quality level of his interactions. Since you have only one person to watch carefully, this task actually takes little energy, and you will still be able to take in the total group interaction.

In the modified fishbowl you will not be able to write down what your partner does, but you can still pay special attention to him. Of course, your first goal is to be a Contributor and not an Observer. Still, try to give your partner some feedback on his interpersonal style ("Three out of your four interactions were questions").

Feedback

Giving feedback is also a skill. The tendency of most participants is to be too long-winded in giving feedback ("an ounce of interaction followed by a pound of analysis"). Another problem is not being concrete enough. *Be brief and concrete. Describe* your partner's behavior, give some indication of your judgment on the quality of his transactions ("Twice you responded to Mary with AE I; I think both responses were accurate, for each time she seemed to pick up and move on further in exploring herself"). If you've had a chance to keep a written record of your partner's transactions, there is no reason why your feedback can't be short, direct, and concrete. It can be supportive in its very objectivity. If your feedback contains some element of confrontation, follow the rules for confronting in giving it. Try to fit your feedback into the total group picture, but, again, resist the temptation to become long-winded and analytic. Twenty minutes of group interaction can be talked about for hours, but not fruitfully.

> *Partner B:* All the members of the group seemed hesitant. Each was waiting for the other to initiate something. You hesitated with the rest, even though in our planning session you seemed eager to get your agenda "on the floor." I think this is part of your style: you initiate well in an initiating group, but a hesitant group binds you up.

If you let concreteness be your guide, along with helpfulness, your feedback will be more useful. The purpose of feedback is to help your partner be more effective in translating his planning into action the next time he is in the group. If your feedback runs into a kind of post-mortem, it is no longer helpful. If your partner does poorly and needs encouragement, provide him with encouragement, not with endless analysis. In the feedback session, *do what is useful.*

In the two-group fishbowl, the As should receive about *five minutes* of concrete feedback. In the modified fishbowl you will need ten minutes, since each partner gives the other feedback at this time.

Switching Roles

In the two-group fishbowl, the whole process is now repeated (planning, group session, observation, feedback) for the Bs. In the modified fishbowl, partners are given a couple more minutes to take another look at their agenda; they then move back into the full group for a second session.

The entire process can be repeated a number of times if this kind of structure is seen as worthwhile. In Phase II, I make extensive use of the fishbowl structure to give the participants ample time to become aware of and make use of group-specific skills. As the participants become more and more proficient in group-specific skills, the amount of structure can be reduced. Finally, the fishbowl itself can be eliminated and the group moves into Phase III, the open-group experience. Below is a schematic summary of the fishbowl.

Two-group fishbowl	*Modified fishbowl*
1. Five or six minutes are spent reviewing agenda with partner (A plans, B helps).	1. Both partners review agendas and help each other.

2. A 20- to 25-minute group session is held, As in inner circle, Bs in outer circle.
3. During session, Bs in outer circle observe partners, keeping notes on their interpersonal style.
4. Bs give As feedback for five minutes.
5. Steps 1-4 are repeated, with Bs in inner circle, As observing.
6. The process is started over again and repeated as long as it remains fruitful (within time limitations).

2. Total group meets for 20-25 minutes.
3. Partners pursue their own agenda, but each partner observes his partner with special care.
4. Partners spend about ten minutes giving each other feedback.
5. After feedback and replanning, total group meets for second group meeting.
6. The process is repeated as often as time allows and the exercise remains fruitful.

At the beginning, the participants are usually self-conscious about the fishbowl format. This self-consciousness may manifest itself in group members' initiating conversations by saying such things as "Well, my agenda is . . . " or "Actually, Tom, *you* are my agenda!" However, such self-consciousness usually diminishes quickly and the structure becomes unobtrusive. One way of proceeding with structure is to move from a two-group fishbowl to a modified fishbowl. But even when the fishbowl format is eliminated entirely and the group moves into Phase III, the high-level communicator will still come to the group with an agenda, in the sense of having concrete objectives he or she wants to accomplish.

The fishbowl format does not of itself make Participants out of Observers or Contributors out of Participants. Therefore it is still necessary to challenge members who, for one reason or another, don't challenge themselves.

CHAPTER 12: FURTHER READINGS

On social influence:

King, S. *Communication and social influence.* Reading, Mass.: Addison-Wesley, 1975.
Rubin, Z. (Ed.). *Doing unto others.* Englewood Cliffs, N. J.: Prentice-Hall, 1974.

PART 6

Phase III: Putting It All Together

The first chapter of this book presented the basic contract for a laboratory experience designed to help the participants examine and begin to change certain aspects of their interpersonal styles. If you have become adept at group-specific skills, in addition to the three sets of basic interpersonal skills described and illustrated in the preceding chapters, you are now ready for an open-group experience—that is, a group experience that, although governed by a basic contract, has comparatively little structure. In the open group, you are trying to improve skills and are no longer merely acquiring them. If you have the basic individual and group skills, you are well on your way to becoming a Contributor in the open group.

In Phase III, you should begin to get a feeling for your own potency as a group member. In other words, you should be aware of your own legitimate needs with respect to the goals of the group and know how to get your needs met in responsible ways through the resources of the group. You should also be aware of the legitimate needs of others in the group and make your resources available to them. For instance, if lack of assertiveness is still a nagging problem for you, the group is a safe place for you to be assertive, and you have a right to expect that your fellow group members will help you place assertiveness demands on yourself. Or, if you see one group member avoiding another, you have a right to ask them both to deal with this problem openly in the group. One way of defining or describing passivity is to see it as a learned inability to get your own needs met or to *do* something about a situation. For instance, if Group Member A claims that she "just doesn't have any impact on anyone in the group," she needs to review her presence in the group from the viewpoint of power, or potency, or influence. If she finds that she

- doesn't respond very frequently to others (is not a source of human nourishment in the group),

- is reluctant to share her feelings (there is an emotional neutrality about her interactions),
- doesn't confront or challenge anyone (is a victim of the MUM effect), or
- finds it difficult to explore her relationship with anyone when she is under any kind of pressure (here-and-now immediacy is not in her repertoire of skills),

she has surrendered four extremely important sources of interpersonal potency. Group members, if they are to have an intensive group experience, have to come to grips with the realities of social influence. It is essential to learn, in a practical, experiential way, that it is all right to have an impact on others and to open yourself up to impact from others. Nonmanipulative social influence is a source of interpersonal potency rather than a surrender of your own independence.

The basic contract outlined in the first chapter of this book gives you a certain number of rights within the group (and also lists your corresponding obligations). If you don't exercise your rights (for instance, the right to experiment in reasonable ways with your interpersonal behavior and to make mistakes—since you do so with the expectation that your fellow group members will both respect you and provide resources to help you correct your mistakes), then it is not reasonable to be resentful of others who do take the initiative to meet their own legitimate needs. It is up to you to personalize the contract, and part of this process of personalizing is making your own needs or wants known to the group. In order to fulfill the contract, you have to "stretch." It isn't meant to be easy. You can't keep your body fit unless you are willing to stretch it in a variety of ways (for instance, with respect to flexibility, muscle capability, and cardiovascular capacity). Similarly, the contract asks you to stretch your interpersonal-communication capabilities.

Chapter 13

The Open Group

DEFENSIVENESS: DEALING CONSTRUCTIVELY WITH FLIGHT BEHAVIOR

It isn't easy to engage in the kinds of interaction outlined in the previous chapters. Therefore, human nature being what it is, we have a natural tendency to find ways either to resist or to flee the work of the group. The message of this section is simple. Don't flee; rather, learn how to deal constructively with flight behavior. The tendency to flee is not necessarily a sign of ill will. Even if you want to commit yourself to this training experience, you will tend to resist the process from time to time, for it is anxiety-arousing and demanding. The group "threatens" you with both self-knowledge (of your interpersonal style) and intimacy (relationship-building), and your defenses rise to the challenge. Flight tendencies appear whenever the human organism is threatened by the often painful processes associated with personal and interpersonal growth.

This section is meant to "blow the cover" of the person (or group) in flight. It is a challenge to the participants to become aware of some of the principal kinds of flight behavior (many have already been alluded to in previous chapters) and to take a stand against flight by confronting it. Flight behavior within the group may mirror the flight behavior in your day-to-day life.

No attempt has been made to classify all possible flight modalities, given the ingeniousness of the human spirit in both devising modes of flight and disguising them. Since not only individuals but entire groups may engage in flight, the following discussion is divided into two sections: the individual in flight and the group in flight. There is, of course, some overlap. For instance, if a number of group participants are engaged in the same kind of individual flight, the group as a whole will also be in flight.

THE INDIVIDUAL IN FLIGHT

Boredom. Boredom, they say, is an insult to yourself. It usually means that you have stopped directing your own life within the group and have come to depend on the will of others. The person who is bored sees himself as a victim of what is happening within the group. His tendency is to put the blame "out there," because the interaction is not "interesting." If you are bored, you are probably acting neither as a Contributor nor as a Participant, but merely as an Observer. When you blame other members in the group for your boredom, you become a Detractor. The excuses dished up by the bored person once he is confronted usually have an anemic cast to them (for, when challenged, the bored person knows instinctively, if only subconsciously, that he deserves his own boredom).

- "I just couldn't get with the discussion."
- "I didn't think the interaction between you two was going anywhere."
- "Nobody was really involving himself in the group."
- "I couldn't seem to get started."

The bored person's nonverbal behavior hangs heavy on the group. Since he is not really attending, he becomes a distraction. Finally, he must be "dealt with" (since he is not dealing with himself), and this is not the most effective use of the members' energies.

What is the remedy for boredom? First of all, *you*, the individual group member, are responsible for your own boredom. Your first tasks are to be aware of the fact that you are slipping away from the group, to identify what is happening inside yourself and within the group to cause your boredom, and to face these issues *immediately*. Boredom, like many other emotional states, only gets worse if it is not faced quickly. For instance, if A and B are being very vague and general as they discuss their relationship, and if in addition no one else in the group is providing any resources to help them be more concrete with each other, *you* can avoid boredom by "owning" or "buying into" their dialogue.

> *You:* Jack, when you say you're uncomfortable with Susan, I'm not sure what you mean. I'm not sure whether you're saying that she expresses affection for you too readily, or that she wants you to respond to her more than you do. I think I have areas of discomfort with both of you; but, before I can contribute anything, it would help me to hear the two of you become more specific.

If your boredom stems from your inability to turn your interests toward the lives of others and away from your own concerns, you might bring up the issue of your self-centeredness as the basis of your boredom. If you wait until you are bored, and then continue to wait until some member of the group calls you on it, you are doing both yourself and your fellow group members a disservice. The best way to handle boredom is to *forestall* it. On the other hand, if the question of someone's boredom does come up (for instance, you challenge someone for inattention and he declares that he is bored), a great deal of the group's time should not be spent on that member. If he is bored, let it be noted, and let *him* or *her* do something about it (move from Observer to Participant or Contributor). If some particular issue underlies the boredom (for instance, the bored person has developed no interest in the group member who has become a focus of the group conversation for a while), someone else should challenge him, or the bored person should challenge himself, to deal with the issue. Avoid long debates on who is responsible for the bored person's boredom. *He* is. Debate in this instance is a diversionary tactic.

Siphoning. If you have some issue that pertains to the group and you deal with it with someone outside the group, you are "siphoning"—that is, draining off energy from the group. For instance, let's say that you and Tom don't get along very well in the group. One evening after a group meeting you decide to see each other for an extended period over the weekend. During this time, you resolve a number of the issues that divide you. You return to the group and say nothing, but now you relate to each other warmly and openly. I don't mean to imply that you should not talk to other group members outside the group. Rather, if you do, you should let your fellow group members know in some summary way what has happened. Otherwise, the lack of continuity in your relationship will be distracting. To take another example, suppose that you and another group member become very close friends outside the group. In the group you may say nothing about this, pretend you are not close friends, or send rather distracting nonverbal messages to each other. Again, something is going on outside the group that affects what takes place within the group. The group does not need a detailed description of how you two relate, but the members deserve enough information to be able to relate to your relationship intelligently. In general, letting others know what has happened between meetings that affects the quality of your participation in the group is sound practice.

Siphoning can also mean that I take counsel with myself about matters that pertain to the group and then keep the conclusions to myself.

> *Group Member A:* I sat down by the lake for a couple of hours last week and just thought about how I act in here with you people. I'm always so agreeable, and I get rewarded for it. It's easy to do, and, if I'm not mistaken, everybody "likes" me. I'm not like that from day to day. I'm often disagreeable. In justice to you and to myself, I think I should act in here more as I do out there. Otherwise it's just too much of a game for me.

If A had just come in one day and had begun to act quite differently, without explanation, others might have found her behavior quite disconcerting.

Psychologizing. The lab is not group therapy (even though the group experience may well be therapeutic), and being overly psychological can be one way of fleeing the real work of the group. There are various ways to psychologize.

- I make myself a helpee instead of a peer of the other group members, a person who is responsible for his own growth;
- I make myself a helper, one responsible for the healing and growth of others, instead of one committed to mutuality; or
- I put undue emphasis on the lab as psychological process, letting its methodology own me instead of serve me. Sometimes a sign of this tendency is language filled with technical terminology, instead of the language humans ordinarily use to communicate.

One common form of psychologizing flight is the interpretation or insight game. The group member who is overly taken with psychodynamic interpretations of and insights into his own or others' behavior is a person in flight. Such interpretations stand in the way of closer human contact.

> *Group Member B:* I think I'm beginning to see why I'm not more assertive here. I still wonder what my parents will think about what I do. Even though my dad is dead.

There is great temptation to look at oneself in terms of interpretation and insight instead of in terms of interpersonal behavior, because the latter carries an implied demand for change through hard work. Looking for new insights into oneself, instead of acting on insights already gained, becomes a way of life. The person who is hoping to be saved by insight is pursuing an illusion; he believes that in the misty future there is an "ultimate insight" that will be an answer to everything. Probably the only way of identifying a "good" insight is to see whether the person who gets it acts on it.

The one kind of insight that does make sense is the kind that results from *doing* something rather than just thinking something.

> *Group Member C:* This morning, Sally, when I told you that I thought you were castrating Fred by treating him so harshly, my heart was going a mile a minute. But I learned two things. First, I learned that I interpret other people's behavior instead of just describing it. Second, I learned how hard it is for me to challenge someone I see as stronger than myself. I'm basically a kind of peace-at-any-price person.

This kind of insight is not a game but a starting point for behavioral change.

The best way to treat your own psychologizing is to ask yourself whether people in the lab (or in real life) are more important than the processes going on. Do I want to make contact with these people? The lab is a kind of experiment in human behavior, but it can be a very human experiment. In order to put a halt to psychologizing, first ask yourself whether you indulge in it. Second, if you see it in another person, *describe* the kind of behavior that you find puts distance between you and that person.

> *Group Member D:* Keith, you practically never explore how you and I relate. When I attempt to do that with you, you listen very attentively, but you tend not to respond except by saying that it's a good point and you should think about it. You *do* interact with me a lot, but since there's so little "you-me" stuff I feel you're always *helping* me. The more I recognize this the more resentful I get.

Here D describes the way Keith flees immediacy by maintaining a helper-helpee relationship and also expresses how he feels about such psychologizing.

Playing the director. It is possible to be very active in a group without moving very deeply into the lives of one's fellow participants. The Director is one who does his best to see that others get involved but, in the process, forgets to get involved himself. The Director specializes in certain kinds of interaction.

> *Group Member E:* Ben and Christina, here's what I see happening when you two interact. Christina plays the quiet, shy little girl. Then you, Ben, come on fatherly, or at least big brotherly. You talk very simply and even slowly to Christina, as to a child. Then, Christina, you get angry because Ben is babying you. I think you're angry right now—but at least to me it has become a kind of little game that you two don't even know you're playing.

The Director buys into ongoing conversations not by sharing himself but by trying to help others relate more effectively. He asks a lot of questions. If there is a lull in the interaction, he comments on it but doesn't seize the initiative to involve himself with anyone in the group. He is very aware of the group process but is not a real part of the process. In many groups he would be seen as a "good group member," one who "gets things moving," for he is "very active" and even has "leadership qualities." But he is a pseudo-Contributor, for he fails to contribute *himself.* He exempts himself from the contract and becomes instead the executive director of the contract. Playing this role is one way that you or a fellow group member can stand back from the process of the group. A Director is prey to all the little ways of avoiding mutuality that plague groups. For instance, it is often a Director (or someone who has fled at least temporarily into a Director-like approach to the group) who will ask the group-cliché question "How do you feel right now?" If he really wanted to get in touch with the feelings of another, he would share his own:

> I'm a bit frustrated right now because I'm not quite sure how you're reacting to me. You're rather quiet, but that doesn't help me very much. I suspect you're angry with me, but I may be wrong.

This person expresses how he feels and how he thinks the other person is feeling. He avoids the Director-like cliché question; he opts for mutuality, and he is more likely to get a meaningful response. The pseudo-Contributor makes demands on others without at the same time making demands on himself; and this is flight, however artfully concealed.

Rationalizations. Rationalization plays a part in almost every form of defensiveness or flight. In laboratory groups, an almost infinite variety of rationalizations are available to the participant who finds that he is not giving himself in a wholehearted way to the group experience. As the proverb goes, he who cannot dance says that the yard is stony. It is possible to mention here only a few of the more commonly used rationalizations. One is the greener-fields syndrome: "I would be doing well if only I were in that other group." It is the peculiar combination of personalities in *this* group that prevents this rationalizer from getting on with the work. Once he has convinced himself that the obstacle is his environment (the group, the trainer, the kind of contract, the room, the furniture, the weather, the time of day) rather than himself, he proportions his participation to his discovery: "Even I can't be effective under these circumstances."

It's too easy to project one's own inadequacies on the laboratory experience itself: "This is a contrived situation, quite unreal; it doesn't facilitate real interpersonal contact." Other statements that summarize rationalization processes are:

- No one else is keeping the contract very well.
- I really don't know what's holding me back.
- I'll let you know whatever you want; just ask me.
- I'm so tired tonight.
- I'm really going to try from now on.
- I really don't know how I can improve my interpersonal relations.

Those who blame the lab structure or any other facet of the lab experience for their failure to get into it or grow ignore the fact that most of their fellow participants *are* getting into the experience and growing through (not in spite of) the structure. Even if some facets of the lab experience seem overstructured to a particular individual (and this certainly does happen), the creative person deals openly with his or her objections to the structure and adapts to what can't be changed. He or she doesn't let structure get in the way.

By now you should be familiar enough with group interaction to make up a list of rationalizations of your own (ones you have used, ones you have heard from others). What is your list like?

Silence. Some have tried to rationalize silence by claiming that there is no evidence that the silent person isn't benefiting from the group experience. A rather silent member of one group suggested an interesting rationalization for his nonparticipation. He said that he was participating *outside* the group but found it impossible to participate *inside* the group. Inside he was *learning;* outside he was *doing.* The silent person may well be learning something in the group, but he is not a Participant or a Contributor—and is therefore violating the group contract. Others defend the silent person by saying that he is a "point of rest" in the group, or that he is "dynamic," since the rest of the members must deal with him. Even if the silent member is dynamic in that he is the starting point for some of the dynamics of the group—his silence is felt by everyone—his silence is still manipulative (at least unconsciously) and counterproductive. The silent member frequently arouses the concern of the group ("Why is he silent?" or "What is happening inside him?"), creates feelings of guilt ("We've been neglecting him"), or provokes anger ("Why does he choose to remain an outsider?" or "Why does he sit there in judgment on the rest of us?"). Silence does manipulate, whether manipulation is the intention of the silent member or not.

A friend of mine told me that once during a 40-hour marathon group experience a girl sat next to the wall for nearly 15 hours, saying nothing. When the facilitator asked her what was troubling her, she said that she had paid her money and was waiting for him to do something with her. This incident, of course, reveals something about the lack of structure of the group experience to begin with. But even in groups governed by a contract calling for initiative, some participants try to take refuge in silence. Although quality of participation is, absolutely speaking, more important than quantity, there is a point at which lack of quantity is damaging to the overall quality of an individual's participation.

The silence of any member should be understood, if possible. Is it fear? Boredom? Lack of interest? Confusion? An ingrained Observer pattern in life? Still, it is counterproductive to use too much of the group's energies in trying to deal with the silent member. Like any other member, he is expected to deal with himself. Otherwise, the silent member in effect demands to be treated like a helpee. This demand calls for therapy rather than a human-relations-training group.

Humor. Humor is a two-edged sword; it can be used to ease the punitive side effects of confrontation and thus facilitate interaction, but it can also be used to obstruct attempts to deepen the level of the interaction or broach sensitive topics. When humor is used to dissipate tension, it does just that, but it does so without getting at the issues underlying the tension. Whenever either an individual or a group adopts humor as a *consistent* component of interactional style (for example, when the group spends five or ten minutes in banter at the beginning of a session, or when a member becomes humorous whenever a particularly sensitive issue is brought up), such behavior needs to be challenged. Once the group has laughed at an interaction to reduce the tension, it is still quite possible to pursue the sensitive topic.

> *Group Member B:* You know, Ned and Phyllis, there is a degree of humor in the way you two interact, but for me at least there is also something sad. You seem to have so much to offer each other, yet you spend so much energy keeping each other at bay.

I don't wish to deny the value of humor in individuals and in groups, however. The humorless individual and the group that can't laugh at its own incongruities are both dull.

Lack of directness. One mode of flight is to speak in generalities rather than specifically and concretely. One way of speaking in generalities is to use "you" and "one" and "we" and "people" instead of the first-person pronoun "I."

> *Group Member A:* People get scared to talk about what bothers them.
>
> *Group Member B:* I'm really afraid to talk about what is bothering me here in the group.

Here B speaks directly and owns his statement, whereas A does not. There is a world of difference between these two statements. High-level communicators seek out ways to be direct instead of using vague language. Directness does make a difference—when you talk about specific incidents and people inside the group instead of speaking generally about the group culture, or when you direct a high percentage of your remarks to individuals within the group instead of speaking generally to the whole group.

In experimenting with groups, I have arbitrarily banned the use of such vague pronouns as "one" and "people" and "you" (when the person speaking means "I"), refused to let general statements go unchallenged, and demanded that the participants address themselves to particular individuals rather than to the entire group. The resulting culture dramatized the lack of immediacy that preceded it, although it seemed initially somewhat artificial (some participants were speechless for a while, discovering that their entire interactional style had a generalized cast to it). Rid the group of the generalities of content and manner that plague it, and you rid it of one of the most common sources of boredom.

Low tolerance for conflict and emotion. Often in groups there are some participants who have a low tolerance for conflict or strong emotion. When conflict and emotion arise, these participants react in one of two ways (or in both ways at different times): they either withdraw from the interaction or try to stop what is taking place. When conflict or confrontation arises, they may become mediators, saviors, or "Red-Crossers." They defend the person being confronted, chide the confronter, and in general pour oil on the waters of conflict. It is not only negative emotion that they find disturbing; strong positive emotions are just as threatening. These, too, must be tempered through humor, change of subject, and other ploys.

I don't suggest here that all conflict is good, or that conflict should somehow be one of the principal characteristics of training groups. However, research (Nye, 1973; Walton, 1969) shows that conflict, if faced reasonably, contributes to rather than detracts from growth. Conflict that is swept under the rug will be constantly tripped over by group participants. If conflict makes you feel uncomfortable, it is best to declare your discomfort without trying to sabotage interactions characterized by conflict.

Conflict is not the same as an unbridled dumping of hostility on fellow group members. Hostility, of course, can express much more than raw "againstness." Especially in group interaction, it can mean many things.

1. It can be a way of expressing one's individuality or showing strength in the group: "I am a potent person." This use of hostility, however, is relatively immature and usually characterizes interactions only in the earlier sessions of the group. Real strength, potency, and individuality can be expressed by becoming a Contributor.

2. For the person who feels threatened by any particular group interaction (for instance, who finds it difficult to engage in self-disclosure of any depth and finds himself in the middle of a group session in which members are questioning the "thinness" of self-disclosure in the group), hostility may be a defensive maneuver rather than a form of attack. Such a person knows by instinct that the best defense is a good offense.

3. Planned hostility ("I've got to do something to stir this group out of its lethargy") can be used as a dynamite technique to stimulate action during a boring session.

4. According to some, hostility can also have a more subtle and constructive meaning: it may be an indirect attempt to achieve some kind of interpersonal contact and intimacy. The hypothesis supported by some research is that identification (or a feeling of closeness) tends to follow aggression. At times two people, after they storm at each other, do tend to draw closer together. Perhaps they find a more direct road to intimacy too difficult, and the turmoil of the indirect route is all that is available.

Hostility as it is described here is a form of flight, for it is doing something indirectly that may well be done better directly. Therefore, hostile interchanges should be looked at carefully—not because hostility is necessarily "evil," and not because a certain amount of hostility isn't normal in human transactions—but because of what the hostility may be covering.

SELF-DEFEATING ROLES

Schwartz (1975) caricatures a number of nonfacilitative roles that trainers fall into from time to time. The roles he describes, however, are not limited to trainers. They are nonproductive ways of assuming leadership, whether by the trainer or by a member. Since the genuine person is role-free (in the sense already described), locking oneself into any particular role is ultimately constricting and phony. Here are a few of these roles.

"The egg beater." This type has to keep things stirred up. He is the group goad, satisfied only when the interaction is at a fever pitch. Although the group is not a place for meditation, neither must it be a frenetic experience. High-level groups are active (the members are working and contributing), but the responsibility for action lies with each member, not just with action-provoking specialists.

"The threshing machine." This person believes that even the most trivial interaction deserves extensive processing. Since processing means analyzing what has happened, it involves extensive use of the past tense.

Processing is a poor substitute for here-and-now interaction. It is not useless, but it becomes dysfunctional and sterile when overused. Each of us in the group could benefit by asking ourselves "Am I interacting or am I processing?" Processing is a good defense against intimacy.

"The cross-eyed surgeon with a meat ax." This person is not satisfied until everyone in the group has exposed his guts. Heavy self-disclosure and heavy emotion must be torn from the participant who refuses to engage in self-surgery.

According to Lieberman, Yalom, and Miles (1973), these "blood-and-guts" interactors cause extensive emotional damage in training groups. A "blood-and-guts" group is a caricature of an intense group. Appropriate self-disclosure and sensible risk-taking in a climate of support can lead to an intense group experience without carnage.

"The Sphinx." This person, whether trainer or group member, keeps a stiff upper lip, showing as little emotion (or even movement) as possible. When the "moment" does arrive and he or she reveals a feeling, the group is seduced into thinking that this moment is sacred. Thus the group is manipulated into giving undue attention to the Sphinx's infrequent utterances.

The Contributor is not always "on" (like the "egg beater"), but he is a *consistent* interactant. Human dialogue is more important than pronouncements. Mutuality, not majesty, is his forte.

"The bullfighter." This person's goal is to bait the other participants until

they rise up in anger. He or she may either use the frontal attack or tease the other into blowing up. Hostility, for the bullfighter, is a primary group value because it is so "honest."

Both avoiding conflict, anger, and hostility and courting them for their own sake are defensive postures. To stir up negative emotionality in such a way adds to the artificiality of the group experience. If group members are working hard at establishing and developing relationships, some abrasiveness can be expected. Conflict always arises; what matters is how creatively it is handled.

"The vacuum cleaner." This member is a pseudoscientist who participates by collecting data on the other participants. He not only figures everyone out but has the data to back up his findings. Such a person emphasizes detailed confrontation, letting support and understanding go by the board.

When attending becomes observation, it is no longer in the service of mutuality. The "vacuum cleaner" or "scientist" manifests a disregard for, if not a fear of, intimacy.

"Lover boy" (or girl). This person tries to make sure that the group runs with rivers of affection so that he or she can get into the swim too. Affectionate intimacy, or fusion, is the primary value, and independence and individualism of any kind are the enemy.

Interpersonal living built on affection alone becomes soggy. Significant human learnings also come from experiencing and working through anger, making decisions, managing conflict, and making reasonable demands of others. Gestalt psychologists encourage us to get in touch with the feeling "polarities" within us. "Each aspect of the person is one part of a duality, even though only one side of the duality may be observable" (Passons, 1975, p. 192). In the "softy" there is also the "toughy," even though the latter aspect remains in the background. Getting in touch with "submerged" poles has a liberating effect:

> The task in resolving the polarity is to aid each part to live its fullest while at the same time making contact with its polar counterpart. This reduces the chance that one part will stay mired in its own impotence (for instance, either the "toughy" or the "softy" in me), hanging on to the status quo. Instead, it is energized into making a vital statement of its own needs and wishes . . . [Polster & Polster, 1973, p. 62].

The specialist in affection or anger is probably not using the fullness of his resources.

Nonproductive roles like these are cues. They say something about the deficits and growth needs of the role-player. What nonproductive roles are you tempted to assume within the group? What other roles do you see operating within your training group?

Exercise 57: Flight: An Exercise in Self-Confrontation

This exercise will help you identify the ways in which you are tempted to flee from an intensive group experience.

1. Review the ways of fleeing described above. Make a list of the ones that pertain to you.
2. Add any kinds of flight that you engage in that aren't included above.
3. Check the one or two that interfere especially with your full participation within the group.
4. Share briefly and concretely with the whole group one or two principal ways you resist full participation within the group.
5. Indicate how you would like to change this behavior. Be as concrete as possible.
6. Indicate how you would like to make use of the resources of the group to help you manage your tendencies to flee.

THE GROUP IN FLIGHT

When the group more or less as a whole is in flight, we are faced with a more serious question. Individuals in flight can be confronted, but when a single individual suggests that the group as a whole is fleeing its work in some way, he may be ignored or even punished. It takes greater courage to confront the group than to confront an individual, but it is also more important to do so. What are the demands of such a confrontation?

Perception. You must be aware of what is happening within the group. If you begin to feel uneasy about some interaction, ask yourself what is going on. For instance, you may discover that three of the seven participants have fallen silent in the last 15 minutes. You can ask yourself what is happening that has caused three of the members to withdraw. If one person falls silent, this may well be an individual problem; but if three members fall silent, it is certainly a group problem.

Description. The confrontation should consist primarily of your description of what is happening (or not happening) within the group. As with one-to-one confrontation, avoid judgmental statements, labeling, name-calling, and the like.

> *You:* I'm feeling uneasy right now. Four of us have fallen silent in the last ten or twelve minutes. It's almost as if we were considering the conversation going on among the other three as private. I'd like to ask myself why I'm staying out of the conversation. Does anyone else feel this way?

You don't assign fault, but you do describe what is going on inside yourself and within the group.

Self-inclusion. Not only do you not judge others, but you take whatever responsibility you should for what is happening within the group. For instance, in the example above you admit that you have fallen silent and that you are responsible for your own silence. Because you include yourself in the problem, you can more easily call on the resources of the group to deal with it.

Invitation. Your description of what is happening usually includes, at least indirectly, an invitation to the other group members to examine what's happening.

Courage. All skills demand a certain degree of courage or assertiveness—willingness to take a risk. Since the risk is higher in this case (one person facing the group, as it were), *the members of the group should discuss this kind of confrontation* and establish its legitimacy. As with any confrontation, the confronter's perception may not be shared by anyone else, but he should have the freedom and support necessary to air his views. Once he finds out that his perceptions are not shared by anyone else, he can explore his own behavior and experiences more intelligently ("What's going on with me to make me feel this way?").

Forms of Group Flight

What, then, are some forms of group flight?

Analysis. One of the favorite forms of group flight, I believe, is analyzing past interactions rather than engaging in here-and-now interactions. Analysis is a form of *group* flight, for it easily becomes part of the group culture. One of those tacit understandings that arise is that it's all right to spend a great deal of time analyzing past interactions. Note the difference between the following two interchanges.

> *Group Member A:* Last week, when you told me that I was withholding a lot of myself here in the group, I got angry. I was saying to myself that I had a right to choose how deeply I would involve myself here. And I was also wondering about you. I mean, it seemed to me that I was getting into the action as much as you were.
>
> *Group Member B:* I kept thinking that you weren't understanding what I was saying. Jack tried to help us get somewhere, but I think we were both getting upset. I didn't know how to resolve the issue, and I was glad when someone changed the subject. I was getting the idea that nobody really knew what we were doing. I got the impression that everyone was relieved when we started on something else.
>
> *Group Member A:* I think I was relieved myself. I was hoping to get some help from the others, but when it became apparent that they were as mixed up about it as we were, I thought there was no use going on—at least then.

This isn't really an interaction but a discussion about interaction. How might this conversation have gone differently?

> *Group Member A:* I got very angry last week when you accused me of noninvolvement in the group, but I did a useless thing—I swallowed my anger. I've been thinking about what you said. I guess I'm not really involved with *you*, and I'm not too sure you're involved with me. I'm afraid of you.
>
> *Group Member B:* Well, I'm afraid of you, too, although I hate admitting it. I wonder what's happening between you and me. If we're afraid of each other, no wonder we're not very involved. I'm afraid of you because you seem to express emotion so easily. When you're hurt here, you say it, and I can read it in your behavior. I'm much colder, and what appears to me to be your emotional freedom unnerves and threatens me, so I steer clear of you.

Analysis is not really interaction; it is a substitute for interaction, a way of taking "time out." Analysis is usually vague, hypothetical, and heady. Past interactions get described over and over again, but this process *has no impact on the here and now.* It is different if you say:

> I was confused last week when you confronted me about talking too much. But I've thought about it. You're right! I try to bury people with words. I keep people at a distance with my language skills. I feel I could do it again, right now.

Analysis inevitably involves a heavy use of the past tense. In fact, as soon as you are aware that the past tense has become predominant in a group conversation, it's time to challenge what is going on. I have already suggested that talking about the there and then can be justified only if it is related to the here and now of the group interaction, to this group of people in *this* situation. Analysis does deal with the (past) interactions of *this* group of people, but it doesn't relate to *this* (here-and-now) situation. The fault lies not with talking about the past but with not relating the past to the present. The past should be brought up to stimulate or focus present interactions.

In my experience, the kind of analysis just described and the processing of exercises are two forms of flight that waste a great deal of the average group's time. I don't mean that a certain amount of analysis (of what is happening right now in the group) and processing of exercises (focused, concrete processing that is time-limited and directed toward improving the quality of upcoming interactions) is not worthwhile and even necessary. But analysis and processing are almost inevitably overdone. Perhaps monitoring the amount of analysis and processing that goes on could even be listed as a group-specific skill. At any rate, the trainer should monitor analysis and processing until this leadership skill becomes diffused among the members of the group.

Focusing on one. Groups very often enter into a tacit decision to spend a good part of a given group session focusing on a single individual. In its extreme form, each member eventually has his or her "turn" as the focal point of the group ("It was my 'turn' today; they spent most of the group meeting on me"). This pattern is a form of flight, for a number of reasons. First of all, giving each member "time" to deal with his or her "problems" tends to turn the group into a helping or therapy group rather than a human-relations-training group. If an extended amount of time is given to an individual, it is almost impossible for that individual not to take on the characteristics of a helpee, and it is almost impossible for the other group members to refrain from indulging in helping behavior. The principal characteristic of a human-relations-training group is *mutuality,* which all but disappears when the group involves itself in a dealing-with-one situation. Once a single individual becomes the focal point of the interaction, a number of members usually fall silent, becoming Observers, because they have "nothing to say" to the participant in the limelight—perhaps because they don't have the same "problems" (or instinctively want to avoid becoming helpers). No structure or process that significantly reduces the involvement of a number of members or serves as an excuse for their noninvolvement (precisely what happens in dealing-with-one situations) should be tolerated in a human-relations-training group.

Second, focusing on one may involve another self-defeating process—an attempt to "finish" an issue at one sitting. The group focuses its energies on one member in order to "get his issues out on the table" so that they can be dealt with. If a given issue is important for a given group member, it can be looked at and *returned to* on several occasions, and care should be taken not to give too much time to the issue on any one occasion. If the tacit assumption is "We'll give each member enough time to work through an issue at one sitting," then Parkinson's Law takes over and the work expands to fill all available time. Let's take a look at an example. Almost without noticing how it happens, John finds that it is his "turn" in the group today; all attention is on him. John's "passivity" comes up. Fellow group members ask questions, tell him things they should have told him weeks ago (focusing on one gives the participants the opportunity to dump out all the items they have been saving up), and more or less ask him to "perform" during the time allotted to him. Since one of John's failings is that he allows others to deal with him instead of dealing with himself, focusing on one merely reinforces John's failing. Besides, John can hardly change on the spot. The work of dealing with John expands to fill the available group time, the meeting ends, and John goes off (having suffered through this rather ritualistic ordeal) unchanged. But since he has "paid his dues" the group will probably not bother him for the next couple of meetings ("After all, we gave him a lot of time last week"). The situation would be much more fruitful if John's passivity were noted early in the life of the group, if group members took some time to understand him from his point of view, if he were given encouragement to use the resources of the group to increase his assertiveness, if John—with the help of the others—were to come up with a plan to increase the quantity and quality of his initiation in the group, and if a short time were spent at each meeting or every other meeting reviewing his progress, how he feels about attempting to become more assertive, how others see his attempts, and how they are reacting to him. It would be much better to give John some "on time" each week than to try to take care of his issues (or any one issue) at one extended sitting.

Third, if one of the principal goals of the group is to have its members establish and develop relationships with one another, spending extended periods of time with one person trying to help him grapple with his "problems" is counterproductive. In a group in which focusing on one has become the norm (by tacit decision), it is difficult to violate this norm. For instance, if you were to bring up concerns similar to John's as a way of involving yourself with John, you might be chided with "I don't think we're finished with John." In a group in which focusing on one has become the norm, it becomes impossible to talk to anyone except the person in the limelight. Talking to another group member is "against the rules." No wonder participants drop back into the role of Observer; their freedom to interact with others has been artificially limited.

There are situations in which giving an individual member more time to deal with a certain issue *is* called for. But I personally prefer group cultures that don't make a *habit* of dealing with a single individual in an extended way. Even the most dramatic aspects of group participants' lives can be dealt with without sacrificing the give-and-take culture of the human-relations-training group. If a member reveals a dramatic dimension of his or her life, and this self-disclosure is relevant to *this* group of people in *this* situation (and it must be, if it's appropriate self-disclosure), he or she needs mutuality and support, not necessarily the kind of extended time that would be accorded a helpee or patient. To assume that this person can best be supported by being made the central actor (or patient) in the group for an extended period of time is unwarranted. One of the most effective forms of support is mutual self-sharing, for it is a way of saying "Thank you for trusting me; I trust you, too" or "Thank you for sharing yourself with me; it gives me the courage to share myself with you."

To presume to "handle" completely either a person or some issue he reveals in a single session is a form of grandiosity. The members of the group are free to return again and again to anything that has been said and to relate it to ongoing group interaction. Remember that the primary goal of the group is to examine your interpersonal style, and that you do this by establishing and developing relationships with your fellow group members. This process will certainly put you in touch with some of the problematic areas in your life. Generally speaking, focusing on one is a process used frequently in problem-centered rather than human-relations-training groups.

Ritualizing the group. Once the structure of Phases I and II have been left behind, the training group can too easily manufacture its own rituals, so that one meeting seems to be a carbon copy of the next. When the group takes on this kind of ritualistic atmosphere in which sameness soothes, directionality is lost. The members of the group go through the motions, but relationships are not deepened and behavior is not changed. Ritualization can happen easily if the members don't come to the meetings *with an agenda.* As a participant, I should want to go somewhere at each meeting. If I have nowhere to go, I may well just drift with the tide of the group. In a ritualistic group culture, not only do the same issues come up over and over again, but they are handled in the same way. For instance, Member X's silence and general lack of involvement are dealt with from time to time, but little is done between ritualistic confrontations to bring X into the interaction. It is almost as if the participants said to themselves from time to time "Since nothing in particular is happening right now, we might as well make a group assault on X again." One way to dramatize ritualization within a group is to replay a videotape of, let's say, session 9 along with a tape of session 12. If the same issues are being dealt with in the same way—that is, if it would be impossible for a stranger to determine which was the earlier session and which the later—it may well be that the group has ritualized itself into a safe little community as a form of flight. A ritualized group culture is a sterile thing, of which boredom, reluctance to come (coming late, beginning the meeting with banter, absences), and poor attending behavior during meetings are the inevitable signs.

Lowest-common-denominatorism. When even one person in a group displays indifference toward the goals of the group, the efficacy of the group is lowered. A training group often tends to move along only as rapidly as its slowest member. The problem of the lagging, delinquent, or deviant member is one that plagues many groups. Effective selection processes and a clear-cut contract can help control deviancy by eliminating the unmotivated person (especially if the contract is freely chosen *before* the group begins and not just imposed once the group has started) and by eliminating the kinds of vagueness and ambiguity in group process that often contribute to the indifference or ill will of the deviant member. The systematic approach of this lab is designed to eliminate vagueness and ambiguity.

The problem of lack of motivation is one of the most difficult to handle in training groups. The unmotivated person or the undermotivated person (the more usual case) is a burden to himself and to the rest of the group. The members of the group should discuss the hypothetical problem of the deviant or undermotivated person before the lab begins. If the possibility is discussed, the delinquent member may have a less retarding effect on the group (that is, if some member does prove to be deviant). Whether the member who refuses to become more than an Observer should be expelled from the group is a moot question, but he or she should certainly not be allowed to absorb the energies of the group.

> *Bill:* Reggie, when are you going to open up here?
>
> *Donna:* Bill, I don't mean to interrupt. Well, I guess I do mean to interrupt. A number of us have initiated interactions with Reggie and have invited him to do the same. I'm not too sure that the best use of our energy is to try to convince members to join the action. If Reggie wants to join what is going on, he is, as far as I'm concerned, welcome at any time. But I don't want to mother him.

Reggie has to provide his own motivation. If the Reggies of the world drop out, that is their choice. Groups are naturally reluctant to see any member leave (even if the member really doesn't want to be there), for the remaining members then have to deal with feelings of guilt and loss. If a member does choose to leave, extended post-mortems can become another form of flight.

Irrelevant serious conversations. Sometimes groups flee the contractual work at hand by engaging in worthwhile discussions that really have no relation to the goals of the group. I once sat in on one session of a human-relations-training group and listened for a while to a rather high-level discussion of some of the most important social issues of the day. Had a total stranger come into the group without knowing its general purpose, he would have thought it was a social-action group taking itself seriously. Indeed, such a conversation would have been most appropriate at another time and place, but within the training group it was a way of fleeing from the real purpose of the group.

Some participants are willing to talk about serious problems that are affecting their lives outside the group but unwilling to face the more immediate issues of relationships with fellow members. College students, for instance, are often quite willing to talk about their problematic relationships with their parents or with those with whom they want to be intimate outside the group. Admittedly, these are serious concerns, but usually they are not here-and-now concerns for this group of people. In some cases, as we have already noted, the problems brought up can be made relevant to the group. In most cases, however, such discussions deaden group interaction in the long run.

Irrelevant serious conversations are usually difficult to challenge, for it seems ignoble to interrupt something that is so worthwhile in itself. It is too easy for the group to make a tacit decision to pursue any serious concern of any participant, even though it is not related to the goals of the group. There are some who would say that such a policy is an indication of the "humanness" and "flexibility" of the group. What happens is that one of the group members shows up in great distress. Almost inevitably, the group session turns into a focusing-on-one helping session, since it seems that the only human thing to do is to put aside the ordinary work of the group and attend to the person in grave need. However, there are other ways of handling such a situation. Often the person in crisis will stay away from the group of his own accord, feeling instinctively that the lab group may not be the best forum for dealing with the crisis. If you do meet some crisis between sessions, you are free to call on the resources of the group *outside* the group's ordinary meeting time. Group members frequently call on fellow group members whom they especially trust to help them through some crisis. In the group, then, they deal with the crisis *insofar as it affects their participation in the work of the group*, but they don't see the whole group as the best forum for dealing with the crisis. Very often people in crisis have to go on living and meeting the commitments of day-to-day life. For instance, a woman tells a man that she no longer wants to see him (or vice versa). Although he is hurt and upset and drained, he still has to work and study and fulfill other commitments. Indeed, learning how to move on responsibly in life in spite of occasional crises is very important. I have heard group members make statements like the one below at the beginning of a group session.

> *Group Member A:* My parents announced to me this week that they're going to get divorced. I knew that they were having their ups and downs, but this is a bolt from the blue. I feel pretty strung out, but I've worked out a lot of my feelings with Bill and Ellen outside the lab. I don't want to go into the whole thing here, but I'm still shaken, and I wanted you to know that I'm not quite myself tonight. But I want to be here and I'll do my best to participate.

The person in crisis can also use this opportunity to examine how he or she handles interpersonal crises, for that certainly is part of one's interpersonal style.

Low intensity. I don't suggest that each group meeting should be a draining experience for every participant, but group interaction *is* work, and you should be reasonably tired at the end of a group meeting. Sometimes, however, you may expend much energy making sure that something does *not* happen. Sometimes groups enter into a tacit agreement as to what is or is not to take place within the group. As Whitaker and Lieberman (1967) note, unchallenged modes of behavior tend to pass into the group culture as laws, and once these laws are made they become very difficult to change or abrogate. These tacit understandings can affect almost every aspect of group life. For instance, tacit understandings or decisions can control:

- *content:* We don't talk about sex here.
- *procedures:* When a person is talking, no one should interrupt him, and he should be given all the time he wants to finish.
- *intensity of interaction:* Hostility should never be expressed directly. It should be discussed only after it has subsided.
- *the provisions of the contract:* Coming late to this group is not an offense.
- *interactional style:* Humor is allowed almost anytime during an interaction.
- *group goals:* It is not necessary for us to forge ourselves into a learning community.

- *leadership:* Only the trainer can do certain things, such as challenge what is happening in the group as a whole.

Tacit understandings or decisions sap the vitality of the group and keep it too safe, for in most human-relations-training groups tacit decisions are conservative decisions, decisions that limit risk-taking. Tacit decisions can begin with clearly expressed statements that go unchallenged. For instance, in a group that was slated to meet once a week for 12 weeks in sessions of three and one-half hours, one of the participants said during the first meeting:

> This setup doesn't really allow us much time to get involved with one another; the time is short and we don't see one another between meetings.

It may well be that this participant was fearful of intimacy and wanted to make sure that not too much in the way of intimacy was demanded of the group. Whatever the case, his statement went unchallenged and kept coming up like a refrain in subsequent meetings. It seems that many of the participants tacitly accepted his analysis of the limitations on the group and spent the following weeks trying to live with the frustration that such limitations entailed. What can be done about tacit decisions that rob the group of intensity?

Exercise 58: Challenging the Group Culture

This exercise is designed to help the members review the norms that have been covertly established within the group and that may be sapping the group of its legitimate intensity.

Directions

Outside the group, write out a list of group norms that have been established tacitly and that, in your opinion, rob the group of some of its vitality. Of course, if you don't find any such norms, don't manufacture any.

Examples

- There is tacit agreement that it is not necessary to provide much primary-level accurate empathy. It strikes me that, as a result, the trust level is mediocre and we are overly wary of one another.
- We are to treat everyone in the group the same way; that is, there is a norm that says that we can't like one person more than another, or that if we do we must express this only outside the group.
- There seems to be a law that we start each group meeting afresh. We don't take up where we left off. As a result, we tend to build up to some serious issues, but they are never faced fully. I would sometimes like to begin by asking what issues we faced only partially last meeting.

These norms can be collated and typed up. In this way, the members will be able to see how frequently any given norm was mentioned. Then the group as a whole can decide how they would like to regain the freedom lost through any particular norm. Doing something about one or two norms that dampen the intensity of the group in significant ways can do much to eliminate other less limiting norms.

One particular tacit decision often has a great impact on the intensity of the group—one referring to self-disclosure. Often there is a group norm that states "Do not risk much at all in terms of self-disclosure, for risk-taking self-disclosure on the part of even one member tends to place demands on all members." The remedy is not to encourage wholesale self-disclosure, for this course would lead to psychological nudity and secret-dropping. Rather, find out what topics are fearful topics for the group, and use this knowledge to lead the group into reasonable risk-taking disclosures.

Exercising 59: Anxiety-Arousing Topics

The function of this exercise is to help individual group members take a look at the quality of their self-disclosure and to ask themselves whether fear or anxiety is preventing them from being more open.

Directions

Outside the lab, write down the topics (your own experiences, feelings, and behaviors) you feel you might be afraid to discuss within the group. List only topics that, if discussed, would move you and the group toward laboratory goals.

Examples

Ways I am ashamed of myself (for example, my physical appearance)

- my need to receive and give affection
- my "neutrality" toward some members of the group
- my need to be understood more fully (in general, the ways I feel vulnerable)
- sexuality
- the differences between my interactional style within the group and outside it

The individual topics can be collated in order to see how frequently any particular topic is mentioned. Then each group member should choose a partner from within the group, reveal one or two areas he or she feels would constitute appropriate self-disclosure, and ask his or her partner to deal with these areas in the full group.

The assumption underlying this exercise is that the group will not be enhanced by more meaningful self-disclosure until the individual members are willing to admit to themselves that there are relevant areas (interpersonal issues and concerns) they are afraid to face. I don't mean to suggest that the group can deal with all possible issues that arise. I do suggest that the group interaction will be enriched if group members don't just let self-disclosure "happen" (or rather, not happen).

Exercise 60: The Emotions We Avoid Expressing

This exercise will help identify the emotions that individuals are not expressing, identify the fears underlying this repression of emotion, and expose the "rules" of nonexpression formulated by individual members so that new, facilitating "rules" can be drawn up.

Directions

This exercise is done in the full group. Each member thinks of some emotions that he or she feels should not be expressed and the reasons why they shouldn't be expressed. All members try to contribute to the "pool." For instance:

Member A: I shouldn't express curiosity about what others have not revealed about themselves, because this is butting into the business of others, which isn't allowed.

Member B: I shouldn't get angry at anyone, because he or she will get angry right back at me and I'll be overwhelmed.

Member C: Don't express affection, and then you won't be rejected.

Someone writes down what is said on a large sheet of paper. This is posted on the wall so that all can see the "rules" that limit interaction within the group.

The sheet can be left up to remind individual members when they are hesitant to express some kind of emotion. The members can discuss whether group interaction is being inhibited by these "rules." It is usually easier to deal with such "rules" once they have been brought to the surface.

FLIGHT VERSUS MAINTAINING ADEQUATE DEFENSES

Laboratory learning experiences, if carried out responsibly and with adequate direction, give the participants a relatively safe opportunity to lower some of their customary defenses in the name of growth. No growth experience, however, should demand that you divest yourself of your defenses entirely. But maintaining adequate defenses (even in the process of lowering them) and flight are two different processes. For instance, with respect to self-disclosure, most of us could afford to experiment with increased self-sharing without going to the extreme of psychological nudity. The latter would indeed make us more vulnerable than we need be. Or most of us could open ourselves up to increased feedback and confrontation without fear of psychic brutalization. You will probably find that for you, as for most people, the more common danger is to not drop your defenses enough to allow the laboratory experience to have its impact on you. The person with crumbling defenses either instinctively refuses to participate in laboratory experiences such as the one described in this book or makes it clear from the beginning of the group experience that he or she has tenuous defenses. In order to avoid lowest-common-denominatorism, selection procedures should screen out those with crumbling defenses, so that reasonably strong demands can be placed on those with adequate defenses. The laboratory is an opportunity for you to "stretch" to reach goals that are obtainable only through hard work. If the laboratory is a "really great experience" for you but places few or inconsequential demands on you, it is probably doing little for you–no more than an exercise program that doesn't significantly increase your pulse rate. Too many laboratory participants end the lab by saying that they wish they had done more or that others had demanded more of them. For instance, participants commonly say that others didn't confront them enough. If you feel you are not confronted enough, it may be because you give cues that say "Don't go too far with me," or because you belong to a group in which a tacit decision has been reached not to confront with any kind of rigor. In both of these cases you can do something about it–if you really want to.

This chapter will end with some further open-group exercises. Although Phase III (dealing with the open group) is characterized by the lessening of or elimination of formal structure, exercises need not be avoided completely. However, care should be taken in Phase III not to substitute formal exercises for individual initiative. It is hoped that, by Phase III, members are well on their way to becoming Contributors. But even in a group of Contributors, exercises can help break up group "log jams" and give greater focus and directionality to the group process. Exercises, then, should not be introduced randomly in order to "stir something up." Such use of them by the leader would be a surrender of leadership and directionality. They should be geared to furthering group goals and adapted to the immediate needs of the particular group. An excessive use of exercises in Phase III is an indication that individual initiative is low.

Exercise 61: Stirring up Immediacy Issues: "Unanswered Questions"

The principal goal of this exercise is to get at immediacy issues ("What's going on between you and me?") and confrontation issues that are being avoided in the group. A secondary goal is to learn experientially how questions are often veiled statements. This exercise can provide a rich "data pool" for further group interaction. It can stimulate a group that finds itself on a plateau.

Directions

Assume that the group has six members (or more), A, B, C, D, E, F. One member (anyone) begins by asking another member a question. For instance, A asks B "Do you like being in this group?"

The member asked (in this case B) *does not answer the question.* Rather, he or she in turn asks someone else a question.

No one may ask a question of the person who has *just* asked him or her a question. For instance, if C asks D a question, D may not turn around and immediately ask C a question. However, D can eventually ask C a question—that is, after being asked a question by A, B, E, or F.

The questions should be relevant to the goals of the group, and the questioning should go on for five to ten minutes. Each member should be making mental notes on the issues that are raised. Sometimes it is even good to have one or two observers, who can keep track of the issues raised and can help recall them later. Usually more data is generated than can be handled by this exercise. "Significant" topics have to be chosen for further group or one-to-one discussion.

After five or ten minutes the group returns to normal interaction. Often the structure of the exercise enables individual members to bring up issues that they have been hesitant to introduce in the open group. Each member should ask himself or herself: What significant confrontation issues were brought up in my regard? What immediacy issues were brought up?

The entire group should deal with issues that relate to the structure and the movement of the group itself. For instance, if several people ask "Do you find this group boring?" the issue of boredom in the group should be faced.

It is most important to *use* the data created by this exercise. Too often, groups bring up excellent issues and then do nothing about them in the subsequent interaction.

Exercise 62: Trust and Immediacy: The "Undisclosed Secret"

If the group is to make progress, trust must continually deepen. The purpose of this exercise is to try to discover "snags" in the trusting process. To entrust yourself to a group of people means that you must, to a degree, entrust yourself to *each of the members*. This exercise helps you review how you stand in terms of trust with each of your fellow group members.

Directions

Think of some secret about yourself that you don't want to disclose to this group of people (for instance, something you don't think would be appropriate to disclose). The less you would like to disclose it the better.

Now, *in your mind's eye,* see yourself trying to disclose your secret *privately* to each of your fellow group members in succession. In your mind's eye see yourself trying (or not being able) to disclose your secret. Try to experience how you would feel with each person. Remember, you never actually reveal your secret to anyone.

Processing can take place in the whole group or in a round robin. Perhaps the round robin is a more efficient way of processing this exercise, but the problem with the round robin is that some important information will remain hidden from the other group members.

Tell each person how you felt in disclosing (or in trying to disclose) your secret to him or her. Describe your feelings and what you imagined to be the other's reactions. Discuss with the other what this means in relation to your ability to trust and to entrust yourself psychologically to him or her. Try to discuss what could take place to deepen trust between you and the other.

Note that a group can provide enough support to enable you to entrust yourself psychologically to the whole group, even though you have some reservations about trusting one or two members.

If you discover that a number of people have some misgivings about entrusting their secrets to you, ask yourself what this means in terms of your behavior in the group. Are you too silent (and therefore seen as untrustworthy)? What can you do to increase your trustworthiness in the group?

Exercise 63: Immediacy: "The Person Who . . . Me Most" and/or "What . . . Me Most about You"

This exercise attempts to provide a structure to get at confrontation and/or immediacy issues that are not being faced. It will also help members of the group identify their interpersonal behavior patterns.

Directions

First of all, a fitting verb is chosen by the trainer or the group members to complete the sentences in the title of this exercise. For instance, "The person who *puzzles* me most in this group is . . . ." Or "What *challenges* me most about you is . . . ." Obviously, a number of different verbs may be chosen—"helps," "puts me off," "stimulates," and so forth.

Decide whether you want to use the first or the second sentence. The first sentence has you single out one person from the group: "The person who puzzles me most in the group is John." The second sentence makes you say something about everyone in the group: "What puzzles me most about you, Mary, is your relationship to Bill. Sometimes you talk to him, but in some meetings you ignore him completely."

Whichever sentence you use, take time to think of the *concrete* reasons for your choice. In sentence one, what in this person's behaviors moves you to choose this person? In sentence two, what precise behaviors puzzle (stimulate, help) you? Describe the other person's behaviors and your reactions to them.

Sentence one

Share your choice in the full group.

Since this kind of sharing usually involves elements of confrontation, be careful to follow the rules of responsible confrontation in what you say.

Don't gang up on any single group member.

Sentence two

Since you have something to say to each group member, your sentences may be shared in the full group or through a round robin.

As you receive feedback, see if you can find any themes running through what is said to you. Try to use the resources of the group to identify patterns of behavior in yourself. You will want to strengthen some patterns, and you will want to modify or eliminate others.

Chapter 14

Changing Your Interpersonal Behavior

In Phases I and II of this laboratory experience, a great deal of your energy has gone into learning skills, both basic interpersonal skills and group-specific skills. Learning or improving these skills certainly constitutes an extremely important form of behavioral change. Skills-building is a positive approach to behavioral change that is almost inevitably very rewarding. Phases I and II also give you the opportunity to examine some of your interpersonal behavior in depth. As you watch your own behavior and receive feedback from others, you perhaps begin to notice some *patterns* of behavior indicating that some change is needed in your interpersonal style. For instance, you notice that you are much more passive and nonassertive in interpersonal situations than you had ever realized. Negatively, there are some things you do as you relate to others that you wish you did *not* do (you are ingratiating and overly compliant; you tend to see others as "better" than yourself). More positively, there are some things you would like to do even though right now you aren't doing them (for instance, you don't let others know your own legitimate wants or needs in interpersonal situations; you don't challenge others, even when you see discrepancies in their behavior and you know they would want to be challenged by you). The question is, then, is there a systematic methodology that will enable you to use the resources of the group to try to rid yourself of some self-defeating pattern of behavior, or to acquire a pattern of behavior that will enhance your interpersonal style? The answer, of course, is yes.

A great deal is presently being written on how to manage, control, and change your own behavior (see Carkhuff, 1973b, 1974b; Thoresen & Mahoney, 1974; Watson & Tharp, 1973; Williams & Long, 1975). This book offers not an exhaustive treatment of self-management techniques (for that, see the books cited), but some hints at a self-change methodology that should enable you to use the resources of the group to help you begin to change some aspects of your interpersonal behavior. This methodology will be directed toward helping you change your own behavior rather than toward your helping someone else change his or her behavior (for the latter, see Egan, 1975b).

CHANGING YOUR INTERPERSONAL BEHAVIOR WITHIN THE GROUP

Here is an overview of the ways in which the lab fosters changes in your interpersonal behavior.

THE PROCESS OF CHANGE

Core interpersonal skills. You are trained in three sets of core interpersonal skills:

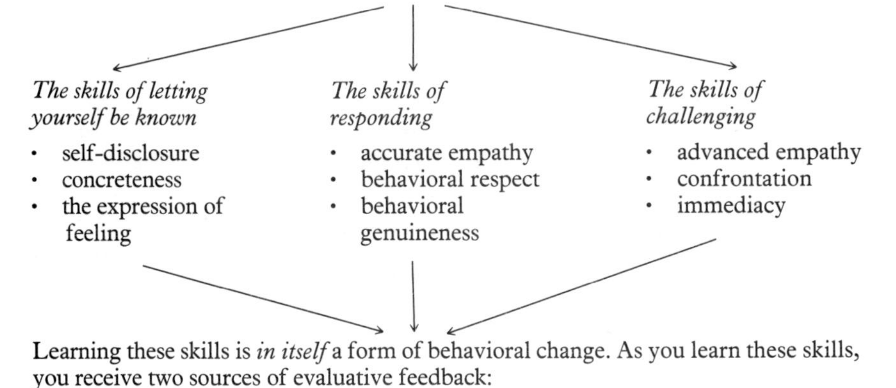

Learning these skills is *in itself* a form of behavioral change. As you learn these skills, you receive two sources of evaluative feedback:

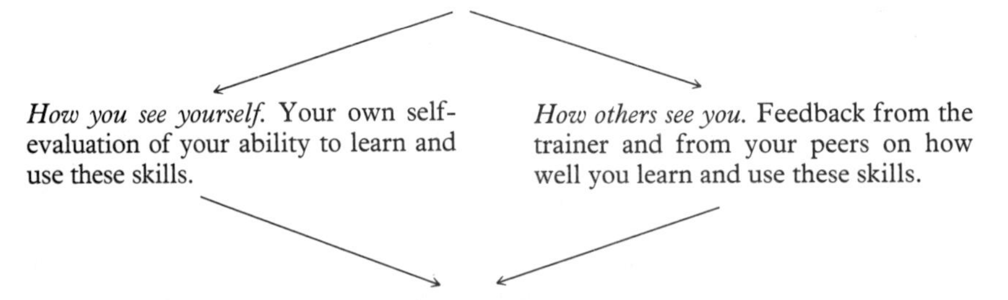

Group-specific skills. You learn how to use these individual skills in a group, and you learn other group-oriented skills:

- how to integrate individual skills into the ongoing process of the group,
- how to "own" into ongoing conversations,
- how to initiate—that is, how to contact others without first being contacted,
- how to ask for feedback, and

- how to use the resources of the group in both giving and responding to confrontation.

↓

Level of initiation. As your ability to use these skills increases, you work toward becoming a Contributor in the group rather than merely a Participant. You begin to have the time and psychological freedom to examine how well you integrate individual and group-specific skills.

↓

Patterns. As you put together both individual and group-specific skills, certain *patterns* of interpersonal behavior emerge:

- the amount you use each interpersonal skill,
- how appropriately you use these skills, and
- how appropriately you integrate and use these skills in order to establish and develop relationships with others.

↓

Owning your unique patterns. You begin to realize that certain identifiable patterns constitute, behaviorally, your interpersonal style. For instance, you discover that

- you seldom use accurate empathy when you are in the group;
- you tend to attend to what others say about themselves, but then move on immediately to your own self-revelation;
- you withdraw when you are confronted, without arguing, without indicating that you have heard the confrontation, and without exploring its content.

That is, you see that you have identifiable *habitual* ways of interacting with others.

↓

Checking out. You find out whether others perceive your interpersonal patterns the same way you do. You try to check and rectify any distortions you may be experiencing. For instance:

- you find that others see your humor as defensive and biting, or
- you discover that you have periods of intense activity in the group alternating with periods of relative silence.

↓

Payoff. You ask yourself and others what payoff there is in your particular interpersonal-behavior patterns. For instance, you discover that

- humor keeps others from taking you seriously. No one expects intimacy from you, and you avoid the work and fear associated with intimacy and closeness.
- periods of intense interaction followed by relatively long periods of silence help you control the outcome of the group, at least with respect to yourself. You are seen as a Contributor because some of your interactions are intense, and you can also choose the times you are going to opt out of group interactions that make you fearful.

↓

Alternative patterns of behavior. You use the group resources to discover alternative, more growthful, patterns of interpersonal behavior. For instance:

- When you enter a "silent" period, you reveal to the other members of the group what is underlying your silence (for instance, fear, because the topic is intimacy in general or sexuality).
- You ask others to make you aware of the times you are using humor, especially if they think it interferes with what is happening in the group or with your making better contact with others.
- You decide to use less self-disclosure and more primary accurate empathy in talking with others, because your interactions are overly self-centered.

↓

Experimentation. You experiment within the group with alternative and more constructive patterns of interpersonal behavior.

Self-evaluation. You evaluate yourself with respect to these new patterns of behavior—that is, how you feel about them (comfortable, awkward), whether they give better expression to the interpersonal values you hold or want to hold, and so on.

Feedback from others. You find out what impact your changed behavioral patterns have on others, how they feel about you and your behavior, how they react behaviorally to the changes, how genuine they see your new behavior, and so on.

Integration and modification. On the basis of your own evaluation and feedback from others, you try to integrate new patterns of behavior into your ongoing style, and/or you continue to try to modify certain patterns of behavior in ways that make interpersonal living more rewarding for both yourself and those with whom you involve yourself.

The group thus provides the context for a great deal of behavioral change. However, if you are to make these changes permanent and transfer them to situations outside the group, certain *principles of change* must get into your bones. Memorizing the following principles is useless; they must find their way into your behavior.

PRINCIPLES UNDERLYING THE CHANGE OF BEHAVIOR

Desire and/or Dissatisfaction

The motivation for change must come from yourself. *You* must want new patterns of acting for the rewards inherent in such patterns, or you must want to change unsatisfactory patterns because *you* are dissatisfied with them. You may get feedback during the course of the group experience that some people don't like certain interpersonal patterns you exhibit (they see you as too aggressive or too compliant or too "heady" in your conversation). But unless you internalize this motivation—that is, unless you make it your own—you probably won't change much of your behavior. In sum, if you yourself are not motivated to change your behavior, any behavioral-change program you undertake will probably fail. There is no magic in the kind of program suggested in these pages, just systematic hard work.

Identification and Clarification

You have to know precisely what behavior you would like to add, subtract, or modify. Your ideas must be very concrete. You can start with a general statement: "I want to be less compliant; I want to be more assertive." But then you must spell out in behavioral terms exactly what "less compliant" and "more assertive" mean. Part of this process involves discriminating assertive behavior from aggressive, hostile, attacking, or manipulative behavior.

- I want to speak up more frequently in this group.
  I want to respond with accurate empathy more frequently.
  I want to challenge group members who say that they want to involve themselves in some contractual behavior and then fail to do so.
- I don't want to manipulate my fellow group members by speaking up just frequently enough to keep them "off my back."
- I notice that it is relatively easy for me to initiate when I'm angry, but then I just tend to spew out my anger unconstructively. I don't want to deny my anger when it arises, but I don't want to use it as a cheap substitute for real initiation.

You are much more likely to change your behavior if what you are trying to change is observable and countable.

> I responded with accurate empathy only once or twice during the last group meeting. Responding with understanding is not instinctive with me at all, even though I am good at practicing this skill in one-to-one situations.
>
> My partner in the fishbowl noticed that I initiated only twice during the last

> meeting. Both times I was angry and, once my anger was out, I fell back into my passive stance.

Concreteness cannot be overstressed. The person who says "I'm shy and I want to do something about it" will do nothing about it unless "shyness" is translated into concrete, specific behaviors.

Accentuating the Positive

Your goal should not be merely to rid yourself of some unwanted behavior. This desire is negative and leads only to a vacuum (and nature abhors a vacuum, we're told). Even if your principal desire is to rid yourself of some negative behavior, this process should be coupled with putting some positive behavior in its place.

> I'm too ingratiating. Whenever anybody in the group says almost anything about himself, I agree with him. I want to be liked and I think people won't like me if I disagree with them. When people speak, I don't want to be an automatic "yes" man.
>
> At first, I want to substitute accurate empathy (AE I) for ingratiating responses. Instead of nodding, saying yes, saying "I understand" and "okay" and things like that, I just want to both understand the other person and give myself time to think.

Accentuating the positive is a way of confronting your strengths instead of just your weaknesses. This doesn't mean that you must be unrealistic and deny your weaknesses, or try to gloss over them with other kinds of behavior. But don't make a behavioral-change process a way of wallowing in your deficiencies. As we will see, this whole process must be rewarding or you won't stay with it long.

Rewarding Yourself

Behaviors that are rewarded tend to be repeated. Behaviors that are punished tend to be repressed. Behaviors that are not rewarded (that go by unnoticed) tend to drop from a person's behavior repertoire. In a word, many human behaviors are governed or controlled by their *consequences.* Therefore, if in your group you take a small, reasonable risk and disclose something about your interpersonal style that you are a bit ashamed of ("I'm a dependent person; I won't initiate anything with you, especially if I see you as strong, until you let me know I'm okay"), and you are rewarded by being understood (AE I) genuinely, you will probably repeat this behavior (take other risks) because, in this instance at least, risk-taking has paid off. Generally, we engage in behaviors in order to attain the consequences we want. If our behavior doesn't help us achieve these consequences, we drop the behavior or switch to something else. For instance, if I see that acting parental toward you (giving you advice, taking care of you, admonishing you) doesn't get you to speak to me or call me up and ask me to do things, I may switch to treating you as an equal.

Let's say that you are interested in becoming more assertive (in whatever concrete ways you define assertiveness for yourself). You won't do so unless there is some payoff for you in becoming more assertive. For instance, you may mistakenly identify being belligerent with being assertive. You become loud and pushy, and all of a sudden you find that your behavior is being punished rather than rewarded. This would be a good time for you to get feedback from your fellow group members.

> Something is going wrong. I've been told that I'm not very assertive at all. I've tried being more assertive, but I may be putting my foot in it. The response has been pretty negative.

Whatever behavior you choose to increase should be a behavior that is normally rewarded (rather than ignored or punished). Within the group, one form of reward is that others pay more attention to what you say and try to include you in their interactions more often. However, if increased personal contact and the intimacy that goes along with it are precisely what you *fear,* you will experience such behavior as *punishment.* A reward or reinforcement is not such unless it is a reward or reinforcement *for you.* If what is rewarding for most people is punishment for you, you must face this problem first.

Reinforcement works best if it takes place *immediately* after the behavior in question occurs (for instance, you communicate accurate empathic understanding to someone and you note how relieved he or she looks or how he or she perks up and reveals more), and if the behavior in question is rewarded *consistently* (for instance, every time—or almost every time—you take time to understand what another person is saying, the person talks more freely, the atmosphere in the group becomes more creatively intense, and you feel good about yourself).

Since reinforcement is so important, you should ask yourself at the beginning of any behavioral-change program you undertake "What is the payoff if I change my behavior?" This is not so crass as it sounds, for payoff is one of the laws governing the maintenance and change of human behavior. Even altruistic acts have some kind of reward or payoff for the person who is altruistic, even though it is not a you-scratch-my-back-and-I'll-scratch-yours reward. The best kind of reward is one that is wedded to the behavior itself. For instance, you may begin practicing interpersonal skills well in order to get a good grade, but you end up learning and practicing them because they are rewarding in themselves. They help create an interpersonal life-style that you find rewarding and satisfying. Finally, reinforcement doesn't work if the behavior you are trying to adopt is not clear or if the steps you have set for yourself are too large. "I think I was assertive today, but not much happened." If you have to say that to yourself, you can't expect much in the way of rewards. The definition of your goal is too vague.

Limiting Your Goals

Once you identify the area in which you would like to work (such as assertiveness), try to establish goals that are limited, possible, concrete, and workable.

I won't try to do everything at once. During the next two sessions, I will try the following:

- I will attend much more carefully to what others are saying. When I'm passive, my body becomes passive, too. I will monitor my posture and my physical involvement with what is happening within the group.
- I will increase the number of times I respond with basic accurate empathy. I find that I have very little contact with Gene and Susan. I will try to respond to these two people especially.
- If I don't get feedback from others during these two sessions, I will elicit it at the beginning of the third session.

Here you specify the behaviors you want to increase. Alternatively, you may want to decrease some behavior (maintaining a slouching posture in the group) or stabilize some behavior (you have spurts of assertiveness and then you sit back passively and become an Observer; people see you as unpredictable and inconsistent).

Baseline Data

In Phases I and II you will learn a great deal about your interpersonal style. In the area in which you want to change, you should have a very clear picture of what you are like *before* you change. For instance, if assertiveness is the area, you should get feedback from others on how they experience you and on how often they see you interacting and taking initiative. Ordinarily you will get this kind of feedback, because of the contractual demands of the group experience itself. Taking note of that feedback, however, is extremely important. The technical term for knowing where you stand with respect to a specific kind of behavior (for instance, communicating primary-level accurate empathy) is "baseline data." You cannot determine whether you are making progress or not unless you have a clear idea of where you started.

> In any given group, I don't initiate more than three times during any session. Of these three instances, usually none is a communication of primary-level accurate empathy.

You know where you stand, and you will be able to determine whether you are making progress. Of course, you don't want merely to increase the number of times you initiate; you are also interested in the quality of your initiations. For information on this, you need to observe how your initiations are received (for example, one of the group members says to you "You've helped me take a more honest look at myself" after you have responded with accurate empathy—an indication that the quality of your initiations is high). You should elicit feedback if it is not given spontaneously.

Boosts and Restraints

You must now take a look at what factors are helping you achieve your clear behavioral goals (boosts) and what factors are keeping you from your goals (restraints). Here again, you must be as concrete as possible.

Restraints

- Whenever I intend to initiate a conversation with someone else, I get nervous. My stomach fills with acid. I perspire. There is a bit of tremor in my legs. Sometimes my voice cracks, and this embarrasses me.
- I'm especially afraid of Nancy. Even when I'm not trying to initiate something with her, I think she is watching me and judging me. I want her to think well of me, so I end up not initiating.
- Sometimes I'm not an accurate perceiver. I communicate what I think is accurate empathy and the person I'm talking to says "No, that's not quite it." I get embarrassed at my failure. I still need much more practice at that skill.

Boosts

- Almost everyone in the group appreciates accurate empathic understanding when he or she receives it. When I do a good job of communicating understanding, I feel good about myself and about what I've done.
- John and I give each other feedback both inside and outside the group. I know he is really "for" me, and he can help me monitor my assertiveness in the group. I trust him a great deal and feel more relaxed when I see him in the group and know that he is "on my side" in a very active way.
- Right now the entire group is trying to increase the amount of empathic understanding that takes place. My effort, then, is part of a larger effort. I won't stand out.

By writing down as concretely as possible your "boosts" and your "restraints," you put all your cards on the table. You are no longer navigating in the dark. Next, ask yourself two questions:

- How can I neutralize or reduce the "restraints"?

- How do I take advantage of the "boosts," especially those that entail the use of group resources (as do two of the suggested three "boosts" above)?

List as many "boosts" and "restraints" as you can think of. Then go over the lists to see what the greatest pitfalls are and what the greatest resources are.

Emotions

If you are doing something that you don't want to do (such as being overly ingratiating and compliant within the group and outside), or if you are *not* doing something you want to (you are not being assertive, you are not challenging your fellow group members to live up to the provisions of the contract), it is quite probable that you are acting or not acting because of some kind of payoff, even though the payoff may be hidden.

Doing what you'd rather not do. Let's say you are ingratiating and compliant, even though you dislike this behavior in yourself. Ask yourself "What's the payoff when I act like this?" You will probably find that there is some kind of emotional payoff, even though you haven't noticed it before. Without being overly interpretative, you may find, for instance, that being compliant rids you of subtle feelings of guilt. You were trained to always be helpful, to obey whoever should be obeyed (and eventually this might mean almost everyone), to respect others carefully, and so forth. In sum, you have learned many ways of acting as you grew up that are tightly linked to certain emotions. You feel self-accepting only if you are meeting the needs of others. Being compliant, then, does have a payoff: you get rid of guilt and you feel good about yourself.

Failing to do what you want to do. Most of us are victims of aversive behavior; that is, we run away from things that are harmful to us. The pity is that we run away from things that we even think might be harmful to us, even though in the long run they would be good for us. Let's take an example. You have a verbal fight with someone. There is a great deal of anger and you both use strong language. Afterward, you begin to avoid each other. You find out that the other person is going to be at a party, so you stay home. You *avoid* the situation in order to *avoid* the negative emotions that would arise if you were to meet the other person. Notice what is happening: each time you avoid the other person, you are, in a kind of backhanded way, being rewarded. Your reward is that you don't suffer the unwanted emotions.

Emotions, then, are very important in a change project. Ask yourself what emotions, both positive and negative, are involved. If you are receiving a kind of phony emotional reward for being compliant, knowing this will help you put aside compliant behavior for more rewarding interdependent behavior. Emotions are tricky; they will deceive you if you let them. As you go over the lists of "boosts" and "restraints," try to find out what emotions are associated with each set. What emotional payoffs keep you wedded to self-defeating behaviors? What emotional payoffs keep you from behaviors that you know will be self-enhancing? The trouble with avoiding (for example, the friend with whom you had the fight) is that you also avoid opportunity for reconciliation. Avoiding usually prevents learning.

Shaping

Don't try to redo your interpersonal life overnight or within a few group sessions. Most self-defeating interpersonal patterns have been learned and reinforced over a number of years. They are "scripts" that you have been playing out for a long time, and therefore they are ingrained. Shaping means that you move step by step toward your goal. You take small steps so that you are rewarded with success as you move step by step toward the goal. Therefore, if you want to increase the number of times you use accurate empathy in the group, you don't go from nothing to everything. You increase your rate gradually and get a feeling of success as you move along. Neither do you try to start with the person you most fear in the group. You start with someone you find easy to talk to. Watson and Tharp (1973) suggest that "will power" is another term for good shaping. In a sense, no step is too small, but any given step can be too large. Shaping also means that you don't start with the most difficult form of behavior. If challenging a person whom you see as very strong and whom you see as not particularly caring for you is one of the most difficult forms of assertiveness for you, don't begin there. Move gradually from the relatively easy to the relatively difficult, feeling your potency grow along the way. Don't underestimate the need for reinforcement. Shaping, then, means starting with an area where some success is both possible and measurable (phoning others, speaking up in class, volunteering, arranging an outing for friends, initiating within the group).

Modeling

Who in the group or in your day-to-day life does well what you would like to do? Watch this person or these persons, not in order to imitate them in any slavish way, but to learn from them, to see how they do it. In the group you will find many ready-made models. Not everyone will do everything well, but each will have his or her own strengths. We have already discussed the group as a modeling network.

A Systematic Program

Don't "just let things happen." The best change programs are systematic programs. Plan ahead. Plan out your agenda from meeting to meeting (your log each week should always conclude with a concrete agenda for the next group meeting). Find someone in the group to help you monitor your change program, and systematically call on the resources of the group (for instance, for feedback). Determine ahead of time what "success" means for you in terms of quantity and quality of behaviors. Evaluate your progress, especially with the help of your "monitor." All systematic change programs have a feeling of artificiality about them. Just as earlier you felt awkward practicing skills you didn't fully possess, you will feel awkward and artificial in engaging wholeheartedly in a change program. In the beginning, the question is not whether you feel awkward or not but whether you are moving toward your goals or not.

Remember, these are only hints to help you become a self-change agent within the group. For a fuller treatment of the issues mentioned here, see the books cited at the end of this chapter.

One final word. As you begin to experience success, *celebrate.*

TRANSFERRING LEARNING FROM THE LABORATORY TO EVERYDAY INTERPERSONAL LIVING

If the interpersonal learning and change outlined above remain locked up in the laboratory itself and don't generalize to your day-to-day interpersonal life, little has been gained. Here let's consider factors that make the transfer of learning difficult and examine ways of getting around these difficulties (see Simpson, no date).

Motivation. Little transfer of learning will take place unless you want it to take place. If you don't want to live more intensively in your day-to-day interpersonal life, the question of transfer of learning is not a real one for you. Unless you are in touch with your motivation (or lack of it), you won't be able to distinguish between "I can't transfer this learning" and "I don't want to transfer this learning." The lab itself is the place to raise the question of transfer of learning. If you never really intend to use the skills you learn in the lab on the outside, your attempts to learn the skills in the first place may be only half-hearted.

Failure to change in the lab. Transfer cannot take place unless you learn the skills well in the lab and begin the process of integrating them into rewarding interpersonal patterns. Poor empathy and poor confrontation are no more acceptable outside the laboratory than within. It is essential to find out whether you are saying "I can't learn these skills well" or "I won't take the time and effort needed to learn these skills well."

Length of the lab. Is the lab long enough and intense enough to enable you to learn interpersonal skills well? Laboratories are work, not magic. A weekend lab may well make you reflect more deeply on the interpersonal issues of your life, but it can hardly be expected to affect long-standing patterns of interpersonal behavior. You need time to gain competence in interpersonal skills and time to integrate them into rewarding interpersonal patterns. The amount of time needed differs from person to person. It isn't easy to realize that others pick up and integrate skills more quickly than you do. If the lab is so constituted that slower learners are actually punished for their slowness, the fault lies in the constitution of the lab itself. However, it is not easy to answer the question "Am I just naturally slower than some others, or am I dragging my feet?" If this is an issue for you, bring it up within the group.

Support for your learning outside the group. It is very difficult to transfer learning to an environment that does not recognize and support your new behavior.

- If intensive living is not a value for those with whom you associate, your new skills won't mean much. You may even be punished for using them.
- If those with whom you associate outside the lab differ greatly from you in terms of skills level and ability to integrate these skills into meaningful interactional patterns, you may once more stand out, and you may be punished for being too far ahead.
- If the environment to which you want to transfer newly learned skills is hostile to these skills, your efforts to humanize interpersonal transactions won't be appreciated. In the business world, for instance, rationality, logic, and getting the task done may be the primary values, and these may take precedence over interpersonal values, especially those involving feelings. If the organization in which you work is authoritarian, impersonal, and characterized by indirect and guarded communication, you can be sure that open communication will be ignored or punished.

This brings up an extremely important point. Your laboratory learning cannot be "inflicted" upon others without their retaliating. Social intelligence, as we have already seen, implies knowing what any particular social situation demands and having the skills and resources to meet those demands. If you merely dump an improved interpersonal style on your friends and coworkers, you can expect them to react.

You have a twofold task before you: (1) integrating what you have learned with your environment in ways that do not alienate and, paradoxically, (2) shaping your environment rather than being shaped by it or victimized by it. One way to shape your environment is to widen your circle of friends and include some that want to grow interpersonally in ways similar to the ways in which you want to grow. Another way to shape your environment is to gradually increase the amount of accurate empathy you use in your transactions with others. Most of us want to be understood and feel rewarded when we are. Understanding others will make possible further attempts at social influence, even confrontation. Shaping means that you move slowly, carefully, and with directionality.

Knowing what you are doing. If the lab you attend lacks directionality and structure, you may well have a good (usually emotional) experience, and there is a tendency to want to repeat this experience "out there." However, since you have experienced little or no *cognitive* input, you really don't know what you did or what others did with you to achieve that "good experience." Part of the learning you should take with you from this lab to your everyday environment is the basic rationale for using each skill and for integrating it in a variety of ways. For instance, if you confront someone outside the lab, you should know that confrontation has to build on understanding; that, at least initially, it should be tentative; and that it should take the form of describing the other's behavior and the impact that this behavior has on you. This kind of knowledge can provide directionality in your interactions with others. It can also lead to the kinds of realization that help you withdraw from unproductive interpersonal behavior. For instance, you might catch yourself confronting another person irresponsibly and say to yourself "I know I've spent little or no time trying to understand this person from his point of view, and I know that I'm angry and merely want to punish him right now. Therefore, it would be better if I could control myself and come back to this subject only when both of us are better prepared for it." Obviously such rationality doesn't always work, but *knowing what you are doing as you are doing it* is a source of power. It makes possible changes that ignorance precludes. At times in interpersonal situations I find myself saying to myself "You're doing practically nothing to communicate understanding to this person." This doesn't necessarily mean that I change my behavior, but I feel that I'm one step ahead in that I know what I'm doing. This lab demands that you learn how to label your interpersonal behavior. This process is awkward and seems to detract from the immediacy of interpersonal transactions. However, the payoff comes outside the lab, where such knowledge can, if you are willing, be translated into more effective behavior.

Moving slowly. The alternative to inflicting a "changed" you on your friends is to let the changes in your style speak slowly for themselves. If others are close enough to you and care about you, you can explain what you have been doing and the importance that these skills now have for you. In this way, you won't make others feel manipulated or left out.

Interpersonal style. What you learn about your interpersonal style as you try to establish and develop relationships with the other members of your group should have some transfer value for your everyday interpersonal life. Even if it doesn't, there is the possibility of learning. For instance, you may notice that you work hard at relationships with your fellow group members but that you fail to do so outside the group. Once you see the difference, you may want to ask yourself "Do I want to pay the price to gain the interpersonal rewards that come from living more intensively outside the group?" Obviously, only you can answer such a question.

I have hypothesized here that skills training transfers much more readily to day-to-day interpersonal living than do the "good experiences" of unstructured labs. Learning skills is a concrete enterprise. It is fairly easy to grasp them both cognitively and experientially. Skills training lessens the possibility of causing laboratory "casualties,' (although as Rogers, 1970, among others, notes, reports of casualties in human-potential labs of various kinds tend to be exaggerated). In laboratory experiences that emphasize skills training first, the trainees are provided the tools they need in order to make their transactions constructive.

THE LIFE-STYLE GROUP

There is one kind of experience that can mediate between what you learn in the laboratory and the demands of your various everyday environments. Once you are grounded in the individual and group-specific skills of the laboratory, you are ready to move on to a group experience that stands outside the laboratory. I call this the "life-style" group.

Plato long ago suggested that the unexamined life is not worth living. Yet most of us go through life without sufficiently examining what we want from it and how we want to give ourselves to it. For instance, it has been suggested that most people choose a career (or fall into a career) and then let the rest of life "fall around" the career as best it can. If I am a victim of this process, my environment owns me instead of my owning my environment. How am I to answer the following questions?

- What do I want from life?
- How do I want to give myself to life?
- What are my real values (the values I actually put into practice), as opposed to my "notional" values (values that right now are only good ideas, since I don't put them into practice in any consistent way)?
- How am I investing myself in life right now?
- How far do my horizons extend? Is my vision limited to my own family? Neighborhood? Town? State? Country? Do I have a "sense of the world"—that is, can I place what I do from day to day in a wider, or even in the widest, context? (For instance, if I live as a middle-class American, do I realize that in the world context I am placed practically at the top of the economic pyramid? And if I am near the top of the world's economic pyramid, is it possible that I have a distorted sense of the world? Isn't it therefore difficult for me to get a realistic "sense of the world"?)
- Is my education just for me, or do educated people in a society have some sort of obligation of service to the wider society and to the educational "have-nots" of that society?
- If I am married or intend to get married, what do I want out of marriage and what do I have to put into it in order to achieve my goals?
- How does my marriage relate to the rest of society?

These questions are, of course, value-laden. In some instances they even imply specific values. In the context of a life-style group, these questions and others like them can be pursued in a practical way—that is, in a way that leads to action. A life-style group is a group of people who, having developed adequate levels of both individual and group-specific skills, use these skills to explore together the kinds of questions and issues listed above. They meet once a week or once every two weeks for fellowship, mutual support, exploration of life-style, and challenge. Their primary goal is not to explore their relationships with one another, but they do review the quality of their relationships from time to time and use the resources of the group to face any interpersonal issues that arise. They are people who have developed high levels of trust with one another; they enter deeply into one another's lives. The life-style group considers such issues as community and social involvement, religion, politics, and the use of leisure, but it does so personally rather than intellectually, practically rather than theoretically. The members are people who want to live more fully and who realize that they must "stretch" to do so. They make use of the "risky shift" phenomenon already described and learn how to become reasonable risk-takers, not just in interpersonal situations but in other areas of life as well. For instance, a member of such a group might ultimately choose to spend a year or two as a volunteer worker in some third-world country as a way of deepening his "sense of the world."

Certainly there is no one rigid definition or description of a life-style group. One group might emphasize the fellowship and challenge aspects of the group, while another might engage in some kind of social action. There are many different ways in which such a group of people can examine and pursue their values, but the pursuit of values is central to the life-style group. I am sometimes uncomfortable with certain aspects of the human-potential movement, for I see an overemphasis on personal and interpersonal growth and relatively little consideration of such things as social issues and development of a sense of the world. Intrapersonal and interpersonal values need not come into conflict with the values of social involvement. I can learn to appreciate meditation and solitude, relate deeply to my friends, earn a living, and still have time for social involvement; but it isn't easy to keep all these values in harmony. One of the functions of a life-style group is to provide the participants a forum for examining the value conflicts that arise in everyone's life and the human resources that enable them to face these conflicts. Whether our society is more heavily cursed with socially destructive individualism and self-centeredness than others is a moot question, but social altruism is not just a human luxury but a necessity. Erikson (1963) has postulated certain developmental tasks that must be faced by individuals at progressive stages in their lives. One of these stages of human development is "generativity." That is, I do not become fully human unless, after seeing to my own needs for identity and intimacy, I make my resources available to others for their well-being, whether these others are my children or other members of society.

Millions of people, of course, must "stretch" just to survive physically. For the most part, intrapersonal, interpersonal, and self-actualization processes are unheard-of luxuries for them. They cannot move higher in Maslow's (1968) self-actualization pyramid because they are crushed at the bottom of the world's economic pyramid. Each of us who numbers himself among the "haves" of this world must ask himself to what degree he can afford the luxuries of this world—material, interpersonal, intrapersonal, or social—in the face of world destitution. At any rate, the life-style group helps its members both develop a more realistic sense of the world and come to grips with the issue of personal responsibility. "No man is an island," wrote John Donne, and in the realm of communications we seem to be fast approaching what McLuhan (1964) has termed the "global village." The life-style group is not a way for the individual to turn in on himself but a way to help him turn toward the world more creatively. To do this effectively, further life skills are needed that are beyond the scope of this book (see Carkhuff, 1973b, 1974a, 1974b, for systematic approaches to the development of some of these skills).

A CONCLUDING NOTE

There is no magic in the pages of this book. There is only the suggestion of hard work. It is impossible to write a book on almost any subject without inadvertently implying that what is being written about is the most important thing in life. Interpersonal skills are extremely important in life, but, important as they are, they must be kept in proportion to the rest of life. Life also involves work, problem-solving, poetry, art, music, involvement in the social order, religion, politics, vacations, romance, learning, fun, books, friends, leisure, pain, career, success, failure, the rest of the world, dying. What you learn in the group should enable you to invest yourself in many other parts of your life more creatively.

CHAPTER 14: FURTHER READINGS

Self-change

Carkhuff, R. R. *The art of problem solving.* Amherst, Mass.: Human Resource Development Press, 1973.

Carkhuff, R. R. *How to help yourself: The art of program development.* Human Resource Development Press, 1974.

Thoresen, C. E., & Mahoney, M. J. *Behavioral self-control.* New York: Holt, Rinehart and Winston, 1974.

Watson, D. L., & Tharp, R. G. *Self-directed behavior: Self-modification for personal adjustment.* Monterey, Calif.: Brooks/Cole, 1973.

Williams, R. L., & Long, J. D. *Toward a self-managed life style.* Boston: Houghton Mifflin, 1975.

Bibliography

Alberti, R. E., & Emmons, M. L. *Your perfect right: A guide to assertive behavior* (2nd ed.). San Luis Obispo, Calif.: Impact, 1974.
Alberti, R. E., & Emmons, M. L. *Stand up, speak out, talk back.* New York: Pocket Books, 1975.
Altman, I., & Taylor, D. A. *Social penetration: The development of interpersonal relationships.* New York: Holt, Rinehart and Winston, 1973.
APA Task Force on Issues of Sexual Bias in Graduate Education. Guideline for nonsexual use of language. *American Psychologist,* 1975, *30,* 682–684.
Aspy, D. N. *Toward a technology for humanizing education.* Champaign, Ill.: Research Press, 1972.
Atkinson, J. (Ed.). *Motives in fantasy, action, and society.* New York: Van Nostrand Reinhold, 1958.
Back, K. W. *Beyond words: The story of sensitivity training and the encounter movement.* New York: Russell Sage Foundation, 1972.
Bayne, R. Does the JSDQ measure authenticity? *Journal of Humanistic Psychology,* 1974, *14*(3), 79–86.
Bebout, J. The use of encounter groups for interpersonal growth. *Interpersonal Development,* 1971/1972, *2,* 91–104.
Bednar, R. L., Melnick, J., & Kaul, T. J. Risk, responsibility, structure. *Journal of Counseling Psychology,* 1974, *21,* 31–37.
Beier, E. *The silent language of psychotherapy.* Chicago: Aldine, 1966.
Benne, K. D., Bradford, L. P., Gibb, J. R., & Lippitt, R. O. (Eds.). *The laboratory method of changing and learning: Theory and application.* Palo Alto, Calif.: Science and Behavior Books, 1975.
Bennett, C. C. What price privacy? *American Psychologist,* 1967, *22,* 371–376.
Berenson, B. G., & Mitchell, K. M. *Confrontation: For better or worse.* Amherst, Mass.: Human Resource Development Press, 1974.
Berne, E. *Games people play.* New York: Grove Press, 1964.
Berrigan, D. *No bars to manhood.* Garden City, N. Y.: Doubleday, 1970.

Berscheid, E., & Walster, E. H. *Interpersonal attraction.* Reading, Mass.: Addison-Wesley, 1969.
Bion, W. R. *Experiences in groups, and other papers.* New York: Basic Books, 1961.
Bradford, L. P., Gibb, J. R., & Benne, K. D. (Eds.). *T-group theory and laboratory method.* New York: Wiley, 1964.
Brammer, L. *The helping relationship: Process and skills.* Englewood Cliffs, N. J.: Prentice-Hall, 1973.
Brandon, N. *The psychology of self-esteem.* New York: Bantam Books, 1969.
Brandon, N. *The disowned self.* New York: Bantam Books, 1971.
Bullmer, K. *The art of empathy: A manual for improving accuracy of interpersonal perception.* New York: Human Sciences Press, 1975.
Carkhuff, R. R. *Helping and human relations: Selection and training* (Vol. 1). New York: Holt, Rinehart and Winston, 1969.
Carkhuff, R. R. *The development of human resources.* New York: Holt, Rinehart and Winston, 1971.
Carkhuff, R. R. *The art of helping.* Amherst, Mass.: Human Resource Development Press, 1972.
Carkhuff, R. R. *The art of helping: An introduction to life skills* (2nd ed.). Amherst, Mass.: Human Resource Development Press, 1973. (a)
Carkhuff, R. R. *The art of problem solving.* Amherst, Mass.: Human Resource Development Press, 1973. (b)
Carkhuff, R. R. *Cry twice!* Amherst, Mass.: Human Resource Development Press, 1974.(a)
Carkhuff, R. R. *How to help yourself: The art of program development.* Amherst, Mass.: Human Resource Development Press, 1974. (b)
Carkhuff, R. R. *The art of helping: Trainer's guide.* Amherst, Mass.: Human Resource Development Press, 1975.
Carkhuff, R. R., & Berenson, B. G. *Beyond counseling and therapy.* New York: Holt, Rinehart and Winston, 1967.
Carkhuff, R. R., & Pierce, R. M. *Teacher as person.* Amherst, Mass.: Human Resource Development Press, 1975.
Carkhuff, R. R., Pierce, R. M., Friel, T. W., & Willis, D. G. *Get a job: The art of placing yourself on a job.* Amherst, Mass.: Human Resource Development Press, 1975.
Carney, C., & McMahon, S. L. The interpersonal contract. In J. W. Pfeiffer & J. E. Jones (Eds.), *The 1974 annual handbook for group facilitators.* La Jolla, Calif.: University Associates, 1974. Pp. 135-138.
Cozby, P. C. Self-disclosure: A literature review. *Psychological Bulletin,* 1973, *79,* 73-91.
Crews, C. C. Use of initial and delayed structure in facilitating group development. Paper presented at the American Psychological Association Convention, New Orleans, 1974.
Culbert, S. A. *The interpersonal process of self-disclosure: It takes two to see one.* Fairfax, Va.: Learning Resources Corp./NTL, 1967.
DeRisi, W. J., & Butz, G. *Writing behavioral contracts.* Champaign, Ill.: Research Press, 1975.
Doster, J. A., & Strickland, B. R. Perceived child-rearing practices and self-disclosure patterns. *Journal of Consulting and Clinical Psychology,* 1969, *33,* 382.

Egan, G. *Encounter: Group processes for interpersonal growth.* Monterey, Calif.: Brooks/Cole, 1970.

Egan, G. Contractual approaches to the modification of behavior in encounter groups. In W. A. Hunt (Ed.), *Human behavior and its control.* Cambridge, Mass.: Schenkman, 1971. Pp. 106-127. (a)

Egan, G. *Encounter groups: Basic readings.* Monterey, Calif.: Brooks/Cole, 1971. (b)

Egan, G. *Face to face: The small-group experience and interpersonal growth.* Monterey, Calif.: Brooks/Cole, 1973. (a)

Egan, G. A two-phase approach to human-relations training. In J. Jones and W. Pfeiffer (Eds.), *The 1973 annual handbook for group facilitators.* San Diego, Calif.: University Associates Press, 1973. (b)

Egan, G. *Exercises in helping skills: A training manual to accompany* The Skilled Helper. Monterey, Calif.: Brooks/Cole, 1975. (a)

Egan, G. *The skilled helper.* Monterey, Calif.: Brooks/Cole, 1975. (b)

Ekman, P., & Friesen, W. V. Nonverbal leakage and clues to deception. *Psychiatry,* 1969, *32,* 88-106.

Erikson, E. H. *Childhood and society* (2nd ed.). New York: Norton, 1963.

Erikson, E. H. *Insight and responsibility.* New York: Norton, 1964.

Fensterheim, H., & Baer, J. *Don't say yes when you want to say no.* New York: Dell, 1975.

Friel, T. W., & Carkhuff, R. R. *The art of developing a career: A helper's guide.* Amherst, Mass.: Human Resource Development Press, 1974. (a)

Friel, T. W., & Carkhuff, R. R. *The art of developing a career: A students' guide.* Amherst, Mass.: Human Resource Development Press, 1974. (b)

Fromm, E. *Man for himself.* New York: Holt, Rinehart and Winston, 1947.

Gazda, G. M. *Human relations development.* Boston: Allyn & Bacon, 1973.

Gergen, K. J. *The psychology of behavior exchange.* Reading, Mass.: Addison-Wesley, 1969.

Gibb, J. R. Defensive communication. *Journal of Communication,* 1961, *11,* 141-148.

Gibb, J. R. Climate for trust formation. In L. P. Bradford, J. R. Gibb, and K. D. Benne (Eds.), *T-group theory and laboratory method.* New York: Wiley, 1964. Pp. 279-301.

Gibb, J. R. The counselor as a role-free person. In C. A. Parker (Ed.), *Counseling theories and counselor education.* Boston: Houghton Mifflin, 1968. Pp. 19-45.

Gibb, J. R. Is help helpful? In W. R. Lassey (Ed.), *Leadership and social change.* La Jolla, Calif.: University Associates, 1971. Pp. 11-17.

Gibb, J. R. Sensitivity training as a medium for personal growth and improved interpersonal relationships. *Interpersonal Development,* 1970, *1*(1), 6-31. (Also published in G. Egan (Ed.), *Encounter groups: Basic Readings.* Monterey, Calif.: Brooks/Cole, 1973. Pp. 311-339.)

Golembiewski, R. T., & Blumberg, A. *Sensitivity training and the laboratory approach: Readings about concepts and applications* (2nd ed.). Itasca, Ill.: Peacock, 1973.

Gordon, T. *Parent effectiveness training.* New York: Wyden, 1970.

Hackney, H. L., & Nye, S. *Counseling strategies and objectives.* Englewood Cliffs, N. J.: Prentice-Hall, 1973.

Hampden-Turner, C. M. An existential "learning theory" and the integration of T-group research. *Journal of Applied Behavioral Science,* 1966, *2,* 367-386.

Hampden-Turner, C. M. *Radical man.* Cambridge, Mass.: Schenkman, 1970.

Harris, T. *I'm OK-you're OK: A practical guide to transactional analysis.* New York: Harper & Row, 1969.

Harvey, O. J., Kelley, H. H., & Shapiro, M. M. Reactions to unfavorable evaluations of the self made by other persons. *Journal of Personality,* 1957, *25,* 393–411.

Henry, J. *Culture against man.* New York: Random House, 1963.

Howard, J. *Please touch: A guided tour of the human potential movement.* New York: McGraw-Hill, 1970.

Huxley, A. *The doors of perception.* New York: Harper & Row (Colophon), 1963. (Originally published: 1954.)

Ivey, A. *Microcounseling: Innovations in interviewing training.* Springfield, Ill.: Charles C Thomas, 1971.

Ivey, A., & Hinkle, J. The transactional classroom. Unpublished paper, University of Massachusetts, Amherst, Mass., 1970.

James, M., & Jongeward, D. *Born to win: Transactional analysis with Gestalt experiments.* Reading, Mass.: Addison-Wesley, 1971.

Johnson, D. W. *Reaching out: Interpersonal effectiveness and self-actualization.* Englewood Cliffs, N. J., 1972.

Jourard, S. M. *Disclosing man to himself.* New York: Van Nostrand Reinhold, 1968.

Jourard, S. M. *Self-disclosure: An experimental analysis of the transparent self.* London: Wiley-Interscience, 1971. (a)

Jourard, S. M. *The transparent self* (rev. ed.). New York: Van Nostrand Reinhold, 1971. (b)

Kagan, N. *Influencing human interaction.* Lansing, Mich.: Michigan State University CCTV, 1971.

Kanter, R. M. *Commitment and community.* Cambridge, Mass.: Harvard University Press, 1972.

Kaufman, G. The meaning of shame: Toward a self-affirming identity. *Journal of Counseling Psychology,* 1974, *21,* 568–574.

Kaul, T. J., & Schmidt, L. Dimensions of interviewer trustworthiness. *Journal of Counseling Psychology,* 1971, *18,* 542–548.

Keen, S., & Fox, A. V. *Telling your story: A guide to who you are and who you can be.* New York: Doubleday, 1973. (Also New American Library, 1974.)

Kelman, H. C. Three processes of social influence. In E. P. Hollander & R. G. Hunt (Eds.), *Current perspectives in social psychology.* New York: Oxford University Press, 1967.

King, S. W. *Communication and social influence.* Reading, Mass.: Addison-Wesley, 1975.

Knapp, M. L. *Nonverbal communication in human interaction.* New York: Holt, Rinehart and Winston, 1972.

Kohn, M. L. Social class and parental values. *American Journal of Sociology,* 1959, *64,* 337–351.

Lawrence, P. R., & Lorsch, J. W. *Organization and environment: Managing differentiation and integration.* Homewood, Ill.: Irwin, 1969.

Lazarus, A., & Fay, A. *I can if I want to.* New York: William Morrow, 1975.

Lee, F. The effect of structure, client risk-taking, and sex of client on early group development. Paper presented at the American Psychological Association Convention, New Orleans, 1974.

Leonard, G. B. Why we need a new sexuality. In G. Egan (Ed.), *Encounter groups: Basic readings.* Monterey, Calif.: Brooks/Cole, 1971. Pp. 260–263. (First pub-

lished in the January 2, 1970, issue of *Look* magazine.)
Levin, E. M., & Kurtz, R. R. Structured and nonstructured human relations training. *Journal of Counseling Psychology,* 1974, *21,* 526–531.
Levitt, E. E. *The psychology of anxiety.* Indianapolis, Ind.: Bobbs-Merrill, 1967.
Levy, R. B. *I can only touch you now.* Englewood Cliffs, N. J.: Prentice-Hall, 1973.
Liberman, R. P., King, L. W., DeRisi, W. J., & McCann, M. *Personal effectiveness: Guiding people to assert themselves and improve their social skills.* Champaign, Ill.: Research Press, 1975.
Lieberman, M. A., Yalom, I. D., & Miles, M. B. *Encounter groups: First facts.* New York: Basic Books, 1973.
London, P. *Behavior control.* New York: Harper & Row, 1969.
Luft, J. *Of human interaction.* Palo Alto, Calif.: National Press Books, 1969.
Lynd, H. M. *On shame and the search for identity.* New York: Science Editions, 1958.
Maddocks, M. In praise of reticence. *Time,* November 23, 1970, p. 50.
Maslow, A. *Toward a psychology of being* (2nd ed.). New York: Van Nostrand-Reinhold, 1968.
Maslow, A. *Motivation and personality.* New York: Harper & Row, 1970.
Mayeroff, M. *On caring.* New York: Perennial Library (Harper & Row), 1971.
McClelland, D. *The achieving society.* New York: Van Nostrand-Reinhold, 1961.
McLuhan, M. *Understanding media: The extensions of man.* New York: McGraw-Hill, 1964.
Mehrabian, A. *Silent messages.* Belmont, Calif.: Wadsworth, 1971.
Melnick, J. Risk, responsibility, and structure: A conceptual framework for initiating group work. Paper presented at the American Psychological Association Convention, New Orleans, 1974.
Mowrer, O. H. *The new group therapy.* New York: Van Nostrand-Reinhold, 1964.
Mowrer, O. H. Loss and recovery of community: A guide to the theory and practice of integrity therapy. In G. M. Gazda (Ed.), *Innovations to group psychotherapy.* Springfield, Ill.: Charles C Thomas, 1968. (a)
Mowrer, O. H. New evidence concerning the nature of psychopathology. *University of Buffalo Studies,* 1968, *4,* 113–193. (b)
Mowrer, O. H. Integrity groups today. In R-R. M. Jurjevich (Ed.), *Direct psychotherapy: Twenty-eight American originals* (Vol. 2). Coral Gables, Florida: University of Miami Press, 1973. (a)
Mowrer, O. H. My philosophy of psychotherapy. *Journal of Contemporary Psychotherapy,* 1973, *6*(1), 35–42. (b)
Nye, R. D. *Conflict among humans.* New York: Springer, 1973.
Passons, W. R. *Gestalt approaches in counseling.* New York: Holt, Rinehart and Winston, 1975.
Polster, E., & Polster, M. *Gestalt therapy integrated: Contours of theory and practice.* New York: Bruner-Mazel, 1973.
Piers, G., & Singer, M. B. *Shame and guilt.* Springfield, Ill.: Charles C Thomas, 1953.
Powell, J. *Why am I afraid to tell you who I am?* Niles, Ill.: Argus Communications, 1969.
Raths, L., & Simon, S. *Values and teaching.* Columbus, Ohio: Charles E. Merrill, 1966.
Roach, A. M. The effects of three ingredients of group structure on early group development. Paper presented at the American Psychological Association Convention, New Orleans, 1974.
Rogers, C. R. *Client-centered therapy.* Boston: Houghton Mifflin, 1951.

Rogers, C. R. *On becoming a person.* Boston: Houghton Mifflin, 1961.
Rogers, C. R. (Ed.). *The therapeutic relationship and its impact.* Madison: University of Wisconsin Press, 1967.
Rogers, C. R. *On encounter groups.* New York: Harper & Row, 1970.
Rogers, C. R. Empathic: An unappreciated way of being. *Counseling Psychologist,* 1975, *5*(2), 2–10.
Rokeach, M. *The nature of human values.* New York: Free Press, 1973.
Rosen, S., & Tesser, A. On reluctance to communicate undesirable information: The MUM effect. *Sociometry,* 1970, *33,* 253-263.
Rosen S., & Tesser, A. Fear of negative evaluation and the reluctance to transmit bad news. *Proceedings of the 79th Annual Convention of the American Psychological Association,* 1971, *6,* 301-302.
Rotter, J. B. Generalized expectancies for interpersonal trust. *American Psychologist,* 1971, *26,* 443–452.
Rubin, Z. (Ed.). *Doing unto others.* Englewood Cliffs, N. J.: Prentice-Hall, 1974.
Ruesch, J., & Prestwood, A. R. Interaction processes and personal codification. *Journal of Personality,* 1950, *18,* 391–430.
Saral, T. B. Cross-cultural generality of communication via facial expressions. *Comparative Group Studies,* 1972, *3,* 473–486.
Schwartz, T. *The responsive chord.* Garden City, N. Y.: Anchor Books (Doubleday), 1974.
Schwartz, T. Trainer styles I have known (and been myself). *The Group Leader's Workshop,* 1975, *21,* 1-4.
Shaffer, J. B. P., & Galinsky, M. D. *Models of group therapy and sensitivity training.* Englewood Cliffs, N. J.: Prentice-Hall, 1974.
Shapiro, S. B. Some aspects of a theory of interpersonal contracts. Psychological Reports, 1968, *22,* 171–183.
Shaw, M. E. *Group dynamics: The psychology of small group behavior.* New York: McGraw-Hill, 1971.
Shostrom, E. *Man, the manipulator.* New York: Bantam Books, 1968.
Simmel, G. The secret and the secret society. In K. Wolff (Ed.), *The sociology of Georg Simmel.* New York: Free Press, 1964.
Simon, S. B. *Meeting yourself halfway: Thirty-one value clarification strategies for daily living.* Niles, Ill.: Argus Communications, 1974.
Simon, S. B., Howe, L. W., & Kirschenbaum, H. *Values clarification: A handbook of practical strategies for teachers and students.* New York: Hart, 1972.
Simpson, C. K. The transfer of learning from the laboratory training group to the outside world. Unpublished paper, Department of Psychology, Cleveland State University, Cleveland, Ohio. (no date)
Simpson, C. K., & Hastings, W. J. *The castle of you: A personal growth workbook.* Dubuque, Iowa: Kendall/Hunt, 1974.
Slater, P. E. *Microcosm: Structural, psychological, and religious evolution in groups.* New York: Wiley, 1966.
Smith, M. J. *When I say no, I feel guilty.* New York: Dial Press, 1975.
Standal, S. The need for positive regard: A contribution to client-centered theory. Unpublished doctoral dissertation, University of Chicago, 1954.
Steiner, C. *Scripts people live.* New York: Grove Press, 1974.
Stern, E. M. Psychotherapy: Reverence for experience. *Journal of Existentialism,* 1966, *6,* 279-287.
Strong, S. R., & Schmidt, L. Trustworthiness and influence in counseling. *Journal of*

Counseling Psychology, 1970, *17,* 197-204.
Strong, S. R., Taylor, R. G., Bratton, J. C., & Loper, R. G. Nonverbal behavior and perceived counselor characteristics. *Journal of Counseling Psychology,* 1971, *18,* 554-561.
Syndnor, G. L., Akridge, R. L., & Parkhill, N. L. *Human relations training: A programmed manual.* Minden, La.: Human Resources Development Institute, 1972.
Talland, G. A., & Clark, D. H. Evaluation of topics in therapy group discussion. *Journal of Clinical Psychology,* 1954, *10,* 131-137.
Tesser, A., & Rosen, S. Similarity of objective fate as a determinant of the reluctance to transmit unpleasant information: The MUM effect. *Journal of Personality and Social Psychology,* 1972, *23,* 46-53.
Tesser, A., Rosen, S., & Batchelor, T. On the reluctance to communicate bad news (the MUM effect): A role play extension. *Journal of Personality,* 1972, *40,* 88-103.
Tesser, A., Rosen, S., & Tesser, M. On the reluctance to communicate undesirable messages (the MUM effect): A field study. *Psychological Reports,* 1971, *29,* 651-654.
Thoresen, C. E., & Mahoney, M. J. *Behavioral self-control.* New York: Holt, Rinehart and Winston, 1974.
Van Kaam, A. *Existential foundations of psychology.* Pittsburg, Pa.: Dusquesne University Press, 1966.
Veysey, L. Individualists bust the commune boom. *Psychology Today,* 1974, *8*(7), 73-78.
Walker, R. E., & Foley, J. M. Social intelligence: Its history and measurement. *Psychological Reports,* 1973, *33,* 839-864.
Wallach, M. A., Kogan, N., & Bem, D. J. Diffusion of responsibility and level of risk taking in groups. *Journal of Abnormal and Social Psychology,* 1964, *68,* 263-274.
Wallen, J. L. Developing effective interpersonal communication. In R. W. Pace, B. D. Peterson, & T. R. Radcliffe (Eds.), *Communicating interpersonally.* Columbus, Ohio: Charles E. Merrill, 1973. Pp. 218-233.
Walton, R. E. *Peacemaking: Confrontations and third-party consultation.* Reading, Mass.: Addison-Wesley, 1969.
Watson, D. L., & Tharp, R. G. *Self-directed behavior: Self-modification for personal adjustment.* Monterey, Calif.: Brooks/Cole, 1973.
Whitaker, D. S., & Lieberman, M. A. *Psychotherapy through group process.* New York: Aldine, 1967.
Williams, R. L., & Long, J. D. *Toward a self-managed life style.* Boston, Mass.: Houghton Mifflin, 1975.
Wilmot, W. W. *Dyadic communication: A transactional perspective.* Reading, Mass.: Addison-Wesley, 1975.
Wood, J. (Ed.). *How do you feel?: A guide to your emotions.* Englewood Cliffs, N. J.: Prentice-Hall, 1974.
Yerkes, F. M., & Dodson, J. D. The relation of strength of stimulus to rapidity of habit-formation. *Journal of Comparative Neurology and Psychology,* 1908, *18,* 459-482.
Zimbardo, P., & Ebbesen, E. B. *Influencing attitudes and changing behavior.* Reading, Mass.: Addison-Wesley, 1970.

Appendix 1
Responses to Exercises

Exercise 13: Rating Risk-Taking in the Expression of Emotion

a–7; b–1; c–5; d–4; e–2; f–8; g–3; h–6

Exercise 25: Identifying Feelings

1. resentful, inferior, angry, put down, in a bind, upset, trapped
2. puzzled, disappointed, gypped, wondering, dissatisfied
3. hurt, rejected, unappreciated, put out, sad
4. grateful, hesitant, afraid, cautious, reluctant, apprehensive, uneasy
5. anxious, uneasy, uncomfortable, awkward, afraid, nervous, frustrated
6. relieved, cautious, satisfied yet searching, settled
7. confused, hurt, left out, resentful, angry
8. uncertain, distant, fearful, hesitant
9. relieved, exuberant, on top of things, successful, assertive, great, good, in good shape, victorious, like he's gotten through

Exercise 26: Possible Responses

1. she feels she has to prove herself to him.
   she's tired of being a second-class citizen.
2. her relationship with Joan seems to be missing something.
   she's been able to get to know only one side of Joan.
3. Bill didn't seem to believe that he would miss him.
   Bill brushed off his concern lightly.
4. he's not sure the group will let him experiment with confrontation.
   he might come on too strong.
5. silences make him uncomfortable.
   silences remind him of his painful times at home.

6. she has found the source of her resentment toward Dottie.
   she realizes she's not angry with Dottie.
7. Frank seems to pay more attention to Mary.
   Frank doesn't seem to care for her as much as he does for Mary.
8. she doesn't want to develop a serious relationship with Sally.
   she's not sure what kind of relationship she wants with Sally.
9. his message has finally gotten through.
   none of them has to play this parent/child game any more.

Exercise 27: Possible Responses

1. ready to explode, angry, furious, ready to wash your hands of the whole group, like pulling out, utterly disappointed
2. at peace, satisfied, fulfilled, like something's being accomplished, like breathing a sigh of relief, like a load has been taken off your back, like you're ready to move closer to each other
3. enthusiastic, like you've found a friend, like cheering, great, on top of the world, very pleasantly surprised
4. confident, relieved, hopeful, pretty good, accepted, in place, safe, secure, like you can risk yourself
5. disappointed, like you have a case of the "blahs," insignificant, depressed, down on yourself, like there's no place to go, hemmed in, sad, down and out
6. grateful, like a new person, good, hopeful, like something good's happening to you, close
7. angry, abused, picked on, resentful, pushed, like blasting him, inferior, put upon, like you have to perform, exasperated
8. wary, cautious, suspicious, not accepted, defensive, on the spot, in the spotlight, like a hold-out, like people are waiting for you
9. eager, brimming over, like moving ahead, enthusiastic, independent, like it's all worthwhile, great, like you don't have to make excuses, full of energy
10. like you've been left hanging, put off, confused, lost, disappointed, rejected, wondering, put out, uneasy

Exercise 28: Possible Responses

1. everybody's so wrapped up in himself and his own needs.
2. you've finally made contact with each other.
3. you never expected to hear such a clear—and digestible—message from me.
4. you're sure that nobody here's going to make a "case" out of you.
5. you just see yourself as too ordinary.
6. your relationship with Mike has so much to offer.
7. you see me just picking away at you without really being "with" you.
8. you almost know that people here are thinking things about you, even though no one has expressed it.
9. you see a lot of opportunities here, and you're determined to take advantage of them.
10. there's such a big difference between our interactions now and what they used to be. You can't help saying "Where have you gone?"

Exercise 29: The Communication of the Understanding of Feelings (More than One Distinct Feeling)

1. protective and frustrated
   accepting and on edge
2. cautious and hopeful
   apprehensive and caring
3. pleased and displeased
   attracted and put off
4. discouraged and scared
   drained and frightened
5. self-satisfied and phony
   satisfied and dissatisfied
6. confused and filled with expectations
   pleased and disappointed
7. eager and rejected
   enthusiastic and betrayed
   hopeful and put down
8. excited and questioning
   pleasantly surprised and hesitant
9. admiring and alarmed
   appreciative and frightened
   grateful and challenged
10. productive and unfulfilled
    competent and frustrated

Exercise 30: The Communication of the Understanding of Content (More Than One Distinct Feeling)

1. You feel both uneasy with Elaine, because your experience has been that women test each other out before taking risks with each other, and hopeful, because you have come to like Elaine very quickly.
2. You feel both attracted because Bill is fun to be with and annoyed because he doesn't seem to know when to be serious.
3. You feel drained, because your friends and others here are having such heavy experiences, and you also feel somewhat scared, because you can't handle any more pain right now.
4. You feel both confident, because you are able to assert yourself here in the group, and helpless, because once outside the group you don't muster up the courage to place demands on others.
5. You feel pleased that John seems to like you and also somewhat disappointed because he doesn't respond to the degree you'd like him to.
6. You feel both eager, because you really like being a facilitator, and rejected, because even your co-trainer has gotten angry with you.
7. You feel excited, because you've never met anyone Kevin's age who is so full of

life, and yet you are a little hesitant, because your view of Kevin demands that you change the way you approach older people.

8. You feel both challenged because Dale takes some real risks here and on guard because you feel inadequate when someone cries.
9. You feel competent and satisfied because you do your work well, but you also feel unfulfilled because you don't have the time to develop your own social life.

Exercise 31: Suggested Responses

1. a. You're feeling angry with me because, by comparing you to Jane and Sue, I zeroed in on one of your more sensitive areas.
   b. By comparing you to Jane and Sue, I've really tapped a vulnerable area. And that upsets you a lot.
2. a. You're feeling frustrated because you're beginning to realize that you keep yourself feeling inadequate by seeing me and others as better than you.
   b. It sounds as if you're really doing a good downer job on yourself—by keeping me and others "one up" and yourself "one down."
3. a. You feel rather isolated and perhaps a bit resentful because, in making something of our helping-profession identity, we leave you out.
   b. It's almost as if you feel out of place here because you're not heading toward some helping profession. In fact, you're beginning to wonder just how different you are from the rest of us.
4. a. You're elated because the group's accepting atmosphere has enabled you to risk more than you ever have before.
   b. It sounds as if it's really refreshing to feel accepted and to even *want* to get down to working at improving your interpersonal style.
5. a. You feel pulled up short a bit, because it may be that all that security and independence is not an unmixed blessing.
   b. Hey, this is something new. It may be that your tough-guy attitude keeps you from some good relationships. That's a tough thing to look at.
6. a. You feel resistant because you don't want to get more involved than you already are.
   b. You're making it pretty clear that you've about reached your limit of involvement, and right now you don't really care to go any further.
7. a. You feel uncertain about how to approach me, because I don't really communicate to you clearly enough how you're coming across to me.
   b. I really don't give you clear-cut messages about how you're affecting me. So you get very cautious and back off.
8. a. You feel in a bind, because you're attracted to me and yet you're frightened by the way I challenge you.
   b. It sounds as if you're not sure which is stronger—the desire to face me or the impulse to run. I turn you on, but I turn you off, too.
9. a. You feel that you just don't know how to relate to me, because I seem so damned "put together" to you.
   b. You're really having a hard time with me. I'm almost too much for you to handle.
10. a. You feel relieved, because you've found a place to explore your sexuality without having to face irresponsible sexual demands or having to act out your own sexual desires.

b. So this is a kind of safe harbor. You can deal with something as touchy and potentially explosive as sexuality directly and seriously, but also responsibly.

Exercise 32: Suggested Responses

1. a. You still feel cautious about getting involved in exercises because they can be unreal, but you also feel good about this one, and maybe any exercise that is reasonable can actually help you be more spontaneous yourself.
   b. It's kind of a pleasant surprise for you to get that much out of an exercise. It doesn't wipe away all of your misgivings, but you seem open to any reasonable way that helps you develop your spontaneity.
2. a. You feel like blasting me because I challenge you in an inept and distracting way, but at the same time you feel bad about yelling at me because you know I care about you.
   b. I'm really driving you up the wall. You know I care, and that part's great, but hounding you is really a lousy way of relating to you.
3. a. You feel bored because we're all being too cautious, but you also feel angry with yourself for not doing anything about it yourself.
   b. You really hate the blandness here, but you also know that you own part of the guilt. You're mad at yourself for letting yourself be bored.
4. a. You feel somewhat sorry that our being together is coming to an end, but you also feel a certain relief because the demands made on you here may be greater than you want.
   b. There's an edge of sadness, at least, in ending. But there's some relief, too. You don't feel that you want interpersonal living to be this intense.
5. a. You feel good because you see that you do affect me at times, but you feel cautious because you are more sensitive—perhaps too sensitive—to what I say to you.
   b. You like it when we deal equally with each other. So when I listen to what you have to say it feels especially good, because there's some anxiety floating around in you about perhaps being a bit dependent on me.
6. a. You feel under some constraint to get through to him, but you also feel reluctant because you can see that it will take a lot of work.
   b. You're in a bind. You'll feel somewhat guilty if you don't live up to the contract, but in this case living up to the contract is very demanding.
7. a. You feel frustrated because your caring doesn't seem to get through to me, but you also feel uncertain because your way of showing care to me may prevent me from seeing it as caring.
   b. You begin wondering whether you're doing this the right way. But the one thing you do know bothers you a lot: I don't respond to what you hope are signs of caring from you.
8. a. You feel a bit wary because this group experience is new, and you know that people have had bad experiences. At the same time, you feel pretty confident because you know people who have benefited by it.
   b. There is some feeling of uneasiness on your part. After all, there are lousy group experiences. But all in all you seem fairly secure. This seems to be the kind of group you can get something from.
9. a. You feel all tied up because all of us seem to take sides, and maybe you also feel hopeless because time is running out.

b. Pairing, including your own, has robbed us of intensity. And that's so much spilt milk. And now you're intimating that maybe it's just too late—or are you?

10. a. You feel great because we pushed through one of our biggest obstacles, our constant dealing with the past. But you also feel on edge because now you have to face your relationship with Bob directly.
    b. You seem almost thrilled that we've overcome that talk-in-the-past problem. But now that we've got time to work, it seems that you've gotten a little more than you bargained for. This thing between you and Bob has thrown you a bit.

Exercise 34: The Identification of Common Mistakes in Phase I

1. a. minus: premature and unfounded AE II
   b. minus: judgmental
   c. minus: premature and poor confrontation; advice-giving
   d. plus: no reason needed
2. a. minus: inaccurate AE I
   b. minus: a nonresponse—no AE I at all
   c. minus: inappropriate question (ignores speaker's feelings)
   d. minus: ignores feelings; poorly worded confrontation; respondent doesn't get into "world" of speaker
3. a. minus: inappropriate respondent self-disclosure; indirect confrontation
   b. minus: inappropriate confrontation; inappropriate AE II; advice-giving; patronizing
   c. plus: no reason needed
   d. minus: inappropriate warmth; side-taking; lack of respect
   e. minus: snide; accusatory; judgmental, unsubstantial AE II
4. a. minus: cliché
   b. minus: patronizing; ignores feelings of speaker
   c. minus: inaccurate AE I (content not accurate)
   d. minus: defensive
5. a. minus: premature and inappropriate use of immediacy ("you-me" talk)
   b. minus: inappropriate question; side-tracking
   c. plus: no reason needed
   d. minus: inappropriate self-disclosure; displaced immediacy
6. a. minus: incomplete AE I (just content); side-taking; approval rather than understanding
   b. minus: inappropriate self-disclosure; judgmental; misplaced confrontation
   c. minus: too long-winded, loses its impact; side-taking
   d. minus: AE I contaminated by misplaced self-disclosure and confrontation
7. a. minus: ignores speaker's statement
   b. minus: picks up only on tangential issue; ignores real feelings and content
   c. plus: no reason needed
   d. minus: ignores feelings and content; respondent pushes his own needs
8. a. minus: defensive
   b. minus: disrespect
   c. plus: no reason needed
   d. minus: inappropriate self-disclosure; avoids issue

Appendix 2
Round-Robin Pairings

A round robin is an opportunity for each member of the group to talk to every other member of the group. First, each member gets a partner. If there is an odd number, the person who is "out" in any round acts as the time-keeper. Thus, in a group of seven, there will be three twosomes talking to each other and one time-keeper. At the end of the specified time, each member moves to a new partner and a new time-keeper is appointed.

In order to save time in choosing new partners, the following pairing system may be used. Each member is assigned a letter. The new pairings are announced at the beginning of each new round. The pairings below are listed according to group size.

Pairings for a 5- or 6-Person Round Robin

1	2	3	4	5
AB	AC	AD	AE	AF
CD	BF	BE	BC	BD
EF	DE	CF	DF	CE

If there are only 5 persons, F's partner sits out.

Pairings for a 7- or 8-Person Round Robin

1	2	3	4	5	6	7
AB	AD	AC	AE	AF	AG	AH
CD	BC	BD	BF	BE	BH	BG
EF	EH	EG	CG	CH	CE	CF
GH	FG	FH	DH	DG	DF	DE

If there are only 7 persons, H's partner sits out.

Pairings for a 9- or 10-Person Round Robin

1	2	3	4	5	6	7	8	9
AB	AC	AE	HI	FG	FH	AG	AH	AJ
CD	EG	BD	GJ	EH	CJ	BH	BF	BG
EF	BI	FJ	AF	AD	DG	DJ	CG	CE
GH	DF	GI	BC	BJ	BE	EI	DI	DH
IJ	HJ	CH	DE	CI	AI	FC	EJ	FI

If there are only 9 persons, J's partner sits out.

Pairings for an 11- or 12-Person Round Robin

1	2	3	4	5	6	7	8	9	10	11
AB	AC	AD	AE	AF	AK	AL	AG	AH	AI	AJ
CD	BD	BC	BF	BE	BL	BK	BH	BG	BJ	BI
EF	EG	EH	CI	CJ	CE	CF	CK	CL	CG	CH
GH	FH	FG	DJ	DI	DF	DE	DL	DK	DH	DG
IJ	IK	IL	GK	GL	GI	GJ	EI	EJ	EK	EL
KL	JL	JK	HL	HK	HJ	HI	FJ	FI	FL	FK

If there are only 11 persons, L's partner sits out.

Name Index

Subject Index